JN300379

化学英語文献への誘い

― 英語演習を通して化学を学ぶ ―

伊藤 浩一・蒲池 幹治 共著

三共出版

まえがき

　本書は，化学および関連分野にいる，あるいはこれから進もうとしている，大学・高専・一般を問わない，初歩の学生・研究者を対象とした化学英語入門書である。卒業研究を始めた学生や大学院で研究に従事する院生には，文献調査，とくに，英語の専門書や論文への取り組みは不可避の作業である。「これだけは知っていれば」，「これだけは理解していれば」，この英語文献をもっと早く，もっと的確に理解できるのだが，という思いは，著者らが長く大学でのゼミや授業で感じていたことである。こんな思いがきっかけとなって，学生や高専生が，化学を学びながら英語を学べるような「化学英語入門書」を目指した。さらに，大学院で研究に携わる院生や若い研究者にも大切で参考となる表現が盛り込まれた「参考書」を志した。

　基礎編では，前半で，初頭に述べた「これだけは」という，化学英語に取り組む場合の基本的事項をまとめ，後半で，いくつかの著名なテキストから引用した基本的演習例題を挙げ，英語の基本文型とともに，化学英語文献読解への手がかりを示した。演習編は，同様なテキストに加えて，最近の2,3の論文も抜粋・引用して，多くの例題を掲げ，和訳例をまとめて別に示した。通じて例題は，おおむね，化学および化学一般，熱力学，分析，無機物質，有機物質，生体関連物質，高分子物質の順に並べたが，基本的な話題のために多分野にまたがる場合も多い。原著からの引用は，原著者の考え・思想を尊重して，原著のままを旨としたが，演習題として独立な内容とするために，ごく一部を著者らの独断で省略・編集した場合もある。また，和訳例もできる限り直訳を旨としたが，文脈によって，かなり意訳も試みた。もし原著者の意向と異なったとしたら，すべて本著者らの責任である。いずれにしても，タイトル毎に，興味深いと思われる内容にまとめることを心がけた。英語演習としてだけでなく，それぞれの「化学」への興味を増幅していただければ幸いである。

　本書は，自習書を基本としたため，すべての例題に和訳例を付した。自習の場合には，まずはできるだけ自分の解釈を試みて，その後で，注，解説，和訳例を参照して，自力の高揚を心がけていただきたい。本書を授業テキストにされる場合は，教官が適当な関連英語文献を課題として補充するのが，学生の英語力・専門知識への意欲を促すと思われる。

　本書の英文とくに例題，演習問題は，多くの著書，テキスト，論文などから引用させていただいた。著者・出版社等に深く感謝を申し上げたい。本書の企画に快く応じていただいた三共出版の秀島功氏にも深謝申し上げる。

　本書が読者諸氏の化学・化学英語への興味を繋ぐ一助となれば，著者この上ない喜びである。

2011年1月　著者記す

凡例：

ex. 例，*cf.* 参照，対比語，(*G*)ギリシャ語由来，(*L*)ラテン語由来
(*n*)名詞，(*v*)動詞，(*vt*)他動詞，(*vi*)自動詞，(*a*)形容詞，(*ad*)副詞，(*prep*)前置詞，(*conj*)接続詞
S＝Subject 主語，V＝Verb 動詞，Vi＝自動詞，Vt＝他動詞，C＝Complement 補語，O＝Object 目的語，［Ex.］Exercise 演習問題

Practice makes perfect!
（習うより慣れよ）

目　　次

基 礎 編
- **1** 覚えておきたい接頭語・接尾語 ·· 1
  - **1**-1 数に関する接頭語　　1
  - **1**-2 有機化合物の炭素鎖，級，分岐に関する接尾語　　3
  - **1**-3 水・油，親・疎(好き嫌い)，分解に関する接頭語・接尾語　　5
  - **1**-4 内外，出入，に関する接頭語　　6
  - **1**-5 反対を表す接頭語　　6
  - **1**-6 その他の接頭語　　7
  - **1**-7 化合物名に関する接尾語　　7
- **2** よく出くわす 化学関連用語 ·· 9
- **3** よく使われる構文 ·· 12
- **4** 基本英文型 ·· 16

演習問題 (Exercise)
- [Ex. 1]　Identifying Molecules and Atoms with Light ·· 15
- [Ex. 2]　Petroleum ·· 16
- [Ex. 3]　General Chemistry ·· 17
- [Ex. 4]　The Early Days of Linus Pauling ·· 22
- [Ex. 5]　Nagasaki 1945 ·· 24
- [Ex. 6]　Host–Guest Chemistry ·· 26
- [Ex. 7]　Sustainable Energy System ··· 27
- [Ex. 8]　Temperature ·· 28
- [Ex. 9]　Equilibrium ·· 29
- [Ex.10]　Hydrogen Bonding ·· 31
- [Ex.11]　Surface-active Agent ·· 32
- [Ex.12]　Combustible Liquid ·· 32
- [Ex.13]　Acid Rain ·· 34
- [Ex.14]　Iron ··· 35
- [Ex.15]　Octane Rating of Gasoline ·· 36
- [Ex.16]　Dimerization of Isobutene to Isooctene ··· 37
- [Ex.17]　Tropolones ··· 38
- [Ex.18]　Chirality ··· 39

[Ex.19] Virus ································ 40
[Ex.20] Carbohydrates ································ 41
[Ex.21] DNA-The Architect of Life ································ 44
[Ex.22] Natural Rubber ································ 49
[Ex.23] Nylon ································ 52
[Ex.24] Great Art in a Test Tube ································ 57

演習編

[Ex.25] Chemistry and Society ································ 61
[Ex.26] SI System ································ 62
[Ex.27] Periodic Law ································ 63
[Ex.28] Radioactivity ································ 64
[Ex.29] Carbon Isotopes ································ 66
[Ex.30] Carbon Dioxide, Henry's Law, and Absolute Zero ································ 66
[Ex.31] Ionic and Covalent Compounds ································ 68
[Ex.32] Energy and Working ································ 70
[Ex.33] Introduction to Chemical Thermodynamics ································ 70
[Ex.34] Objectives of Chemical Thermodynamics ································ 71
[Ex.35] The First, Second, and Third Laws of Thermodynamics ································ 72
[Ex.36] Number and Choice of Components ································ 74
[Ex.37] The Tubular-Flow Reactor ································ 74
[Ex.38] Soap as an Amphiphilic Compound ································ 76
[Ex.39] Origin of Surface Tension ································ 76
[Ex.40] Electromagnetic Radiation ································ 76
[Ex.41] Ultraviolet (UV) and Visible Spectroscopy ································ 79
[Ex.42] Infrared (IR) Spectroscopy ································ 80
[Ex.43] Magnetic Resonance Imaging (MRI) ································ 81
[Ex.44] Metals ································ 82
[Ex.45] Antacids ································ 84
[Ex.46] Clathrates, Methane and Carbon Dioxide Hydrates ································ 85
[Ex.47] Silicon, Silica, and Zeolites ································ 87
[Ex.48] Ions in Your Body ································ 89
[Ex.49] Acid and Base ································ 91
[Ex.50] Superacids ································ 93
[Ex.51] Nature of Organic Molecules ································ 94
[Ex.52] Aspirin ································ 95

[Ex.53]	Sweeteners	96
[Ex.54]	Terpenes	98
[Ex.55]	Drugs	100
[Ex.56]	Alkaloids	102
[Ex.57]	Connecting Biomass and Petroleum Processing with a Chemical Bridge	104
[Ex.58]	Amino Acids to Proteins	106
[Ex.59]	Lipids	107
[Ex.60]	Digestion of Carbohydrates	110
[Ex.61]	Nucleic Acids and Heredity	112
[Ex.62]	Polyethylene and Polypropylene	113
[Ex.63]	Rubber Elasticity	114
[Ex.64]	Silicones	115
[Ex.65]	Implanted Polymers for Drug Delivery	116
[Ex.66]	Biopolymers versus Synthetic Polymers	117
[Ex.67]	Films, Membranes, and Coatings	118
[Ex.68]	Room-Temperature Ionic Liquids	118
[Ex.69]	Block Copolymer Micelles	120
[Ex.70]	Serendipity	120
[Ex.71]	Serendipity in Polyethylene and Polypropylene	121

あとがき 123
引用文献 124
和訳例 [Ex.25〜71] 126

INDEX 159

基 礎 編

Some important tips that will help your learning English in chemistry
(君の化学英語学習を手助けするいくつかの大切な手引き)

　今や英語は国際語である。化学に限らないが，科学関係の英語論文から，世界をリードした，あるいはリードしようとしている新しい知見・概念・考えに接してこれを理解したい応用したいというのは，誰もが持つ思いであろう。卑近な例で言えば，大学や高専でのゼミあるいは授業での学生の思いも，教官の学生諸君への期待もそうである。もちろん研究者自身，教官自身の思いもそうである。自分の研究分野に直接関わる専門論文の理解は，最も差し迫った問題であろう。外国語の論文・論説に接する場合に，「これだけは知っていれば」，「これだけは理解していれば」という基本的な素養の修得が必要である。ここでは，化学の英語論文に対処するときに必要な基本的事項を解説する。「これだけ頭に入れれば」化学英語論文への理解と，そのスピードが加速するのでは，と期待したい事項である。

1　覚えておきたい　接頭語・接尾語

　多くの言語は，単語の知識が豊富であればあるほど，理解の助けになることはいうまでもない。化学英語で使われる用語，単語も，基本的な接頭語，接尾語の複合から成る場合がかなり多いので，これを覚えれば，単語理解力・記憶力が大幅に増すこと必定であろう。

1-1　数に関する接頭語

　およそ，科学論文に接する場合に，数字に関する接頭語・単位に関する用語は，必須の基本事項である。まず，化学でよく用いられる，極小から極大(10^{-12}〜10^{12})の分数・倍数を表す接頭語は，国際単位(SI)系で表1のとおりに定められている。

　これらの接頭語に，例えば，メートル(m)が続けば，見慣れた長さを表す量になる。通常の原子や分子の大きさ，数〜数百オングストローム(Å)，は10^{-10}〜10^{-8} m すなわち0.1〜10 ナノメートル(nm)のオーダーにある。よく言われる「ナノテク」は"nano-technology"のことで，原子，分子の次元で材料設計・機能を目指す技術のことを表している。大きい方では，周波数のメガヘルツ(MHz)，ギガヘルツ(GHz)，コンピュー

表1 SI系で用いられる接頭語例 ( )内は記号

10^{-12}	pico-(p)ピコ		10	deca-(da)デカ
10^{-9}	nano-(n)ナノ		10^2	hecto-(h)ヘクト
10^{-6}	micro-(μ)マイクロ		10^3	kilo-(k)キロ
10^{-3}	milli-(m)ミリ		10^6	mega-(M)メガ
10^{-2}	centi-(c)センチ		10^9	giga-(G)ギガ
10^{-1}	deci-(d)デシ		10^{12}	tera-(T)テラ

表2 1〜10の接頭語と化学関連用語の例

(G) ギリシャ語由来, (L) ラテン語由来

1	mono-(G)	monomer モノマー, 単量体, monosaccharide 単糖, monochrome 単色, monolayer 単分子層
	uni-(L)	unimolecular 単分子の, univalent 1価の
2	di-(G)	dimer 2量体, divalent 2価の, disaccharide 2糖, dihedral 2面の
	bi-(L)	bimolecular 2分子の, biphenyl ビフェニル, bilayer 2分子層, binary 2成分の, bidentate 2座配位の
	bis-(L)	a, a'-azobis(isobutyronitrile) アゾビスイソブチロニトリル: NC-C(CH$_3$)$_2$-N=N-C(CH$_3$)$_2$-CN
3	tri-(G)	trimer 3量体, trivalent 3価の, triangle 3角形, trigonal 3角形の, 3方晶の, triclinic 3斜晶の
	ter-(L)	ternary 3成分の, terdentate ligand 3座配位子
	tris-(G)	tris(p-aminophenyl)methane トリス(p-アミノフェニル)メタン: CH(-C$_6$H$_4$-NH$_2$)$_3$
4	tetra-(G)	tetramer 4量体, tetrahedron 4面体-, tetragonal 正方晶の, 四角形の
	quadr-(L)	quadrangle 4角形, quadrupole 4重極子
	tetrakis-	tetrakis(triphenylphosphine)nickel(0) テトラキス(トリフェニルホスフィン)ニッケル(0): Ni(PPh$_3$)$_4$
5	penta-(G)	pentagon 5角形, pentahedron 5面体
6	hexa-(G)	hexagon 6角形, 6方晶, hexahedron 6面体
7	hepta-(G)	heptagon 7角形, heptahedron 7面体
8	octa-(G)	octagon 8角形, 8方晶, octahedron 8面体
9	nona-(L)	nonagon 9角形
10	deca-(L)	decagon 10角形, decahedron 10面体

ターなど記憶媒体のギガバイト(GB)などが馴染みであろう。

接頭語ではないが, 微量, 痕跡量を表すのに, とくに分析, 環境化学などの分野でよく使われる次の略号も覚えておくべきである:

　　　ppm(part per million):100万分の1 (10^{-6})
　　　ppb(part per billion) :10億分の1 (10^{-9})

数字1〜10にまつわる接頭語は, とくに科学ではよく用いられる。その接頭語と化学に関係する例を表2にまとめた。

数字は特定されないが, 次の多少を表す接頭語もよく用いられる:

　　　oligo- 少数(数個)の　　　oligomer オリゴマー(寡量体), oligosaccharide, オリゴ糖(寡糖)

poly- 多数の	polymer ポリマー（重合体，高分子），polysaccharide 多糖
multi- 多数（数個）の	multicomponent 多成分の，multilayer 多分子層，multifunctional 多官能性の

1-2　有機化合物の炭素鎖，級，分岐に関する接頭語

多くの有機化合物は，炭素数を表す接頭語と化合物のタイプあるいは官能基を表す接尾語の組み合わせから成るので，この基本的なパターンあるいはルールを理解・記憶すれば，化合物名と構造の相関がすぐつくようになる。表3には，炭素数 C_1〜C_5 に関する接頭語と接尾語から成る基本的な脂肪族化合物の例をあげた。

表3 炭素数 C_1〜C_5 に関する接頭語と接尾語（脂肪族化合物の例）

	Typical compounds						
	alkane アルカン	alkyl アルキル	alkene アルケン	alkyne アルキン	alkanol アルカノール	alkanal アルカナール	alkanoic acid アルカン酸
一般式 C_n prefix	C_nH_{2n+2}	C_nH_{2n+1}	C_nH_{2n}	C_nH_{2n-2}	$C_nH_{2n+1}OH$	$C_{n-1}H_{2n-1}CHO$	$C_{n-1}H_{2n-1}COOH$
C_1 meth-	methane メタン	methyl メチル	−	−	methanol メタノール	methanal メタナール	methanoic acid メタン酸
					(methyl alcohol)（メチルアルコール）	(formaldehyde)（ホルムアルデヒド）	(formic acid)（ギ酸）
C_2 eth-	ethane エタン	ethyl エチル	ethene エテン	ethyne エチン	ethanol エタノール	ethanal エタナール	ethanoic acid エタン酸
			(ethylene)（エチレン）	(acetylene)（アセチレン）	(ethyl alcohol)（エチルアルコール）	(acetaldehyde)（アセトアルデヒド）	(acetic acid)（酢酸）
C_3 prop-	propane プロパン	propyl プロピル	propene プロペン	propyne プロピン	propanol プロパノール	propanal プロパナール	propanoic acid プロパン酸
			(propylene)（プロピレン）	(methylacetylene)（メチルアセチレン）	(propyl alcohol)（プロピルアルコール）	(propionaldehyde)（プロピオンアルデヒド）	(propionic acid)（プロピオン酸）
C_4 but-	butane ブタン	butyl ブチル	butene ブテン	butyne ブチン	butanol ブタノール	butanal ブタナール	butanoic acid ブタン酸
			(butylene)（ブチレン）		(butyl alcohol)（ブチルアルコール）	(butyraldehyde)（ブチルアルデヒド）	(butyric acid)（酪酸）
C_5 pent-	pentane ペンタン	pentyl ペンチル	pentene ペンテン	pentyne ペンチン	pentanol ペンタノール	pentanal ペンタナール	pentanoic acid ペンタン酸
						(valeraldehyde)	(valeric acid)（吉草酸）

各枠の中の上段は，IUPAC(International Union of Pure and Applied Chemistry 国際純粋および応用化学連合)の定めた命名，下段のカッコ内は慣用名である．とくに，酸やアルデヒドは，天然由来化合物に使われてきた慣用名が，現在も多く使われていることに注意されたい．

この表を超える炭素数の化合物も，次の接頭語をもとに，容易に対応する命名(とくに，IUPAC 名)が可能であろう．例えば，

C_6 hex-　　　C_{11} undec-　　　C_{20} eicos-
C_7 hept-　　　C_{12} dodec-　　　C_{22} docos-
C_8 oct-　　　C_{14} tetradec-
C_9 non-　　　C_{16} hexadec-
C_{10} dec-　　　C_{18} octadec-

脂肪族環状化合物は，接頭語 cyclo- を付して表される．例えば，cyclohexane シクロヘキサン，cyclopentane シクロペンタンは，それぞれ6員環，5員環を表す．芳香族化合物は，ベンゼン，ナフタレンなど慣用名が基本になるので，基幹化合物名と構造を覚えることが肝要である．

有機化学では，炭素鎖の級または置換度，分岐を表すための接頭語も慣用的によく用いられる．表4に例を示す．その脚注とともに，参考にされたい．

表4 炭素の級・置換度，分岐を表す接頭語

n-(normal 正)[*1]	n-butanol: $CH_3CH_2CH_2CH_2$-OH	n-butyl cation: $CH_3CH_2CH_2CH_2^+$
iso-[*2]	isobutanol $(CH_3)_2CHCH_2$-OH	isobutyl cation: $(CH_3)_2CHCH_2^+$
sec-(secondary 第2級)	sec-butanol $(C_2H_5)(CH_3)$CH-OH	sec-butyl cation $(C_2H_5)(CH_3)CH^+$
tert-(tertiary 第3級)	tert-butanol $(CH_3)_3$C-OH	tert-butyl cation $(CH_3)_3C^+$

注：[*1] n- は付けない場合も多い：例えば，単に butanol, octanol などは n-butanol, n-octanol を表す．
　　[*2] isos (G) に由来するが，「分子量は同じで，違うもの(異性)」を意味する．表の例(イソブタノール，イソブチルカチオンのように，iso (イソ) は実際上化合物名の中に含まれる．他に，isooctane = 2,2,4-trimethylpentane: $(CH_3)_3CCH_2CH(CH_3)_2$ などがある．

参考　(1) primary(第1級)は接頭語として，使われることは少ない．
　　　　上記の例で言えば，butanol も butyl cation も，n- と iso- は primary である．
　　(2) 炭素とアミンの級数(置換度)を表すのに，primary, secondary, tertiary, quarenary が使われることも多い．

	炭　素	ア　ミ　ン
prim-(primary　第1)	R–CH$_2$–X, CH$_3$–X	R–NH$_2$, NH$_3$
sec-(secondary　第2)	R$_2$CH–X	R$_2$NH
tert-(tertiary　第3)	R$_3$C–X	R$_3$N
quart-(quarernary　第4)	R$_4$C	R$_4$N$^+$（第4アンモニウムイオン）

(X は官能基，R はアルキル基（必ずしも同じアルキル基ではない））

注：methane（メタン：CH$_4$）の置換度は，methyl（メチル：CH$_3$），methylene（メチレン：CH$_2$），methine（メチン：CH）で表されることが多い．

(3) 2つの官能基が，1つの炭素につく場合（*gem-* = geminal ジェミナル），2つの隣り合う炭素につく場合（*vic-* = vicinal ビシナル）の接頭語もある（geminal：一対の，vicinal 隣の）：
　　ex.　*gem*-dichloride（CH$_2$Cl$_2$），*vic*-diol（HOCH$_2$CH$_2$OH）

(4) 二重結合を含む不飽和化合物の幾何異性体は，接頭語 *cis-* シス（または *Z-* ツザンメン：ドイツ語の zusammen（一緒に）から）と *trans-* トランス（または *E-* エントゲーゲン：ドイツ語の entgegen（反対に）から）で表される：

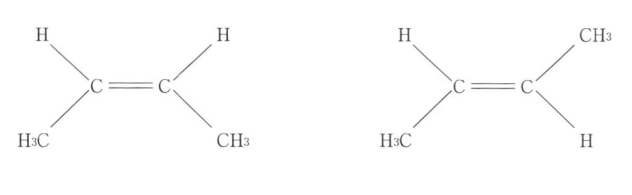

cis-Butene（*Z*-Butene）　　　　　*trans*-Butene（*E*-Butene）

1-3　水・油，親・疎（好き嫌い），分解に関する接頭語・接尾語

およそ，単語・用語の成り立ちは，接頭語・接尾語の組み合わせである場合が多くあり，同義語，派生語，反対語などは，語彙を増やす有効な方法である．以後の数節は，このような化学関連単語を，言わば「芋づる」式に列挙したものである．読者の記憶を助けるとすれば，幸いである．

hydro-　　水の
lipo-　　　油の
-philic　　好む，親-性の
-phobic　　嫌う，疎-性の
　　ex.　hydrophilic 親水性の，hydrophobic 疎水性の，lipophilic 親油性の，lipophobic 疎油性の，nucleophilic 求核性の，electrophilic 求電子性の，nucleophile 求核種，electrophile 求電子種，hydrolysis 加水分解
　　　　cf. philosophy 哲学（学・知を愛する）は，接頭に philo- がくる．

-hydrate　水化物，水和物
　　ex.　carbohydrate 炭水化物，magnesium chloride hexahydrate 塩化マグネシウム・6水和物 $MgCl_2 \cdot 6H_2O$
　　　　　cf. ちなみに，低分子の炭水化物：糖(sugar, saccharide)，さらに他の栄養素：lipid 脂肪，脂質，protein タンパク質

hydro-　水素の
　　ex.　hydrocarbon 炭化水素，hydrochloric acid 塩化水素酸(HCl)

-lysis　－分解
　　ex.　hydrolysis 加水分解，alcoholysis 加アルコール分解，pyrolysis, thermolysis 熱分解，electrolysis 電気分解

homo-　同種の
　　ex.　homogeneous 均一系の，homolysis ホモリシス($A-B \rightarrow A \cdot + \cdot B$)

hetero-　異種の
　　ex.　heterogeneous 不均一系の，heterocyclic 複素環の，heterolysis ヘテロリシス($A-B \rightarrow A^+ + :B^-$)

-ase　－アーゼ(酵素語尾)
　　ex.　lipase リパーゼ(脂肪分解酵素)，hydrolase ヒドロラーゼ(加水分解酵素)，amylase アミラーゼ〈炭水化物消化酵素〉，proteinase プロテナーゼ(タンパク質分解酵素)，polymerase ポリメラーゼ〈重合酵素〉

1-4　内外，出入，に関する接頭語

endo-　中に　endothermic 吸熱の，endocrine 内分泌の
exo-　外に　exothermic 発熱の，exocyclic 環外の
intra-　内の　intramolecular 分子内の，intracellular 細胞内の
inter-　間の　intermolecular 分子間の，
　　　　　　international 国間の(国際の)　cf. national 国(内)の，
　　　　　　interaction 相互作用，interdisciplinary 学際の，intermediate 中間体，
　　　　　　interface 界面　cf. surface 表面，bulk 本体，内部
extra-　外の　extraneous 余分の，extraordinary 異常な，extrapolate 外挿する，

1-5　反対を表す接頭語

a-　　asymmetric 不斉の(=chiral)，cf. achiral 不斉でない
an-　　anhydride 無水物，anhydrous 無水の，anaerobic 嫌気性の，anisotropic 異方性の
anti-　antibonding 反結合の，anticlockwise 反時計回りに，antibody 抗体，antigen 抗原

de-		deactivate 不活性化する，dehydration 脱水(和)，deform 変形する，denature 変性する
dis-		dissymmetric 反対称の，discharge 放電，discoloration 脱色，disappear 消失する
non-		nonbonding 非結合の，nonpolar 非極性の，nonlinear 非線形の，nonmetal 非金属，nonvolatile 不揮発性の
un-		unambiguous 曖昧でない，uncertain 不確定の，unexpected 予期しない，unstable 不安定な，unreliable 不確実な

1-6 その他の接頭語

petro-	石油の	petrochemicals 石油化学製品
thermo-	熱の	thermometer 温度計，thermoplastic 熱可塑性の，thermodynamics 熱力学
trans-	交換の	transesterification エステル交換，transalkylation アルキル交換，transcription 転写，translation 翻訳，transplantation 移植，transport 輸送
ultra-	超-	ultraviolet 紫外の，ultrasonic 超音波の，ultracentrifuge (vt, n) 超遠心分離(する)
per-	過-	peroxide 過酸化物，perchloric acid 過塩素酸($HClO_4$)
iso-	同じ	isothermal 等温の，isobar 等圧の，isoelectric point 等電点，isotope 同位体，isomer 異性体(同じ分子式で，構造の異なるもの)
co-	共に	copolymer 共重合体，coenzyme 補酵素，coplanar 同一平面の，covalent 共有結合の

1-7 化合物名に関する接尾語

1-1にあげた「数」の接頭語に続く多くは，その数の対象となる内容を表す接尾語である。例えば，表2から分かるように，-mer(繰り返し単位)，-saccharide(糖)，-layer(分子層)，-gon(角)，-hedron(面)など自明であろう。また表3から，有機化合物のタイプを表す接尾語，-ane, -yl, -ene, -yne, -ol, -al, -oic acid は，それぞれ，アルカン，アルキル，アルケン，アルキン，アルコール，アルデヒド，酸に対応する。

　無機化合物の命名では，正イオン，負イオンの順に並べるのが原則である。負イオンは，-ide が接尾語(語尾)になる。金属正イオンは，1, 2価の低原子価金属は -ium が語尾となる場合が多いが，複数原子価を持つ金属の慣用名は，-ous(低原子価)，-ic(高原子価)の語尾で区別される。例をあげよう。

-ide －イド（負イオン）または －化物

carbide 炭化物 *ex.* calcium carbide（CaC$_2$）カーバイド，silicon carbide 炭化珪素（SiC カーボランダム）

oxide 酸化物 *ex.* carbon(di)oxide（CO$_2$）二酸化炭素

sulfide 硫化物 *ex.* carbon(di)sulfide（CS$_2$）二硫化炭素

hydride 水素化物，水素アニオン（H$^-$）
  ex. calcium hydride （CaH$_2$）水素化カルシウム

hydroxide 水酸化物，水酸化イオン（OH$^-$）
  ex. sodium hydroxide（NaOH）水酸化ナトリウム

halide ハロゲン化物，ハロゲンイオン（X$^-$）
  ex. potassium chloride（KCl）塩化カリウム，chloride（Cl$^-$）塩化物イオン

（金属）-ous と -ic：低原子価金属（イオン）と 高原子価金属（イオン）

  ex. ferrous chloride FeCl$_2$, ferric chloride FeCl$_3$
    cuprous chloride CuCl（Cu$_2$Cl$_2$），cupric chloride CuCl$_2$

酸素，窒素，硫黄化合物が，非共有電子対にプロトンあるいはアルキルカチオンなどを付加して生じる正イオン（カチオン）は，-onium（オニウム）が接尾語になる。

  ex. hydronium ヒドロニウム（H$_3$O$^+$）
    oxonium オキソニウム（>O$^+$-）
    ammonium アンモニウム（>N$^+$<）
    sulfonium スルホニウム（>S$^+$-）
      cf. carbonium = carbocation カルボニウム＝カルボカチオン（>C$^+$-）
        carbanion カルバニオン（>C$^-$-）

酸，塩，エステル類も，語尾で酸化度などが区別される。
（非金属）-ic と -ous（接尾語）：正（非金属）酸と亜（非金属）酸
    -ate と -ite（接尾語）：正（非金属）塩と亜（非金属）塩

  ex. sulfuric acid 硫酸 H$_2$SO$_4$
    sulfurous acid 亜硫酸 H$_2$SO$_3$
    nitric acid 硝酸 HNO$_3$
    nitrous acid 亜硝酸 HNO$_2$
    sodium sulfate 硫酸ナトリウム Na$_2$SO$_4$
    sodium sulfite 亜硫酸ナトリウム Na$_2$SO$_3$
    potassium nitrate 硝酸カリウム KNO$_3$
    potassium nitrite 亜硝酸カリウム KNO$_2$

（非金属）-ate と -ite（接尾語）：正（非金属）酸エステルと亜（非金属）酸エステル
  ex. dimethyl sulfate 硫酸ジメチル（CH$_3$）$_2$SO$_4$（通称ジメチル硫酸）

    dimethyl sulfite 亜硫酸ジメチル $(CH_3)_2SO_3$
    methyl acetate 酢酸メチル CH_3OCOCH_3
    sodium dodecyl sulfate ドデシル硫酸ナトリウム $C_{12}H_{25}\text{-}OSO_3Na$
    sodium alkylbenzenesulfonate アルキルベンゼンスルフォン酸ナトリウム $R\text{-}C_6H_4\text{-}SO_3Na$ (R = alkyl)

2 よく出くわす 化学関連用語

　化学関連用語を挙げだすと，きりがなく「辞書」，「事典」を参照いただくことにして，ここでは，よく出くわす用語，間違い・混乱しやすいと思われる用語を，「芋づる」式に並べることにする。

化学一般用語

chemical (a) 化学の，chemical (n) (通常複数 chemicals) 化学薬品，reagent 試薬，-agent －剤
　　ex. chemical reaction 化学反応，petrochemicals 石油化学製品，surface-active agent 界面(表面)活性剤

compound 化合物，mixture 混合物，blend ブレンド，composite 複合体，複合材，aqueous (*aq.*) 水の (水溶液の)，
　　ex. *aq.* HCl 塩酸，*aq.* NaOH 水酸化ナトリウム水溶液

adsorb 吸着する，adsorption 吸着
absorb 吸収する，absorption 吸収
sorb 収着する，sorption 収着
　　ex. adsorbed water 吸着水，absorbed light 吸収光，sorbed gas 収着気体
　　　　cf. superabsorbent polymer 超高吸水性ポリマー

kinetics (n) 速度論(動力学)，kinetic (a) 速度論的(動力学的)
　　ex. kinetic constant 速度定数，kinetic order 動力学的次数，kinetic control 速度論的支配
　　　　cf. kinetic energy 運動エネルギー

thermodynamics (n) 熱力学，thermodynamic (a) 熱力学的の
　　ex. thermodynamic control 熱力学支配

qualitative 定性の，quantitative 定量の
　　　　cf. quality 質，quantity 量，quantum 量子

rare metal レアメタル，希少金属，rare earth element 希土類元素 (レアアース)，noble metal 貴金属

rare gas ＝ noble gas 希ガス＝貴ガス，inert gas 不活性ガス

conformation コンフォメーション (立体配座)

configuration コンフィギュレーション (立体配置，立体構造)

化学反応用語

addition 付加, elimination 脱離, abstraction 引抜き, condensation 縮合, substitution 置換, oxidation 酸化, reduction 還元, hydrogenation 水素化(水素付加), hydration 水付加(水和), dehydrogenation 脱水素, dehydration 脱水(脱水和), neutralization 中和

 ex. $HCl + NaOH \rightarrow NaCl + H_2O$
    cf. ←(逆反応):hydrolysis 加水分解

esterification エステル化

 ex. $CH_3COOH + C_2H_5OH \rightarrow CH_3COOC_2H_5 + H_2O$
     cf. ←(逆反応):hydrolysis 加水分解

isomerization 異性化, rearrangement 転位, transfer 移動, shift 移動, hydrogen transfer 水素移動, proton transfer プロトン(H^+)移動, hydride transfer ハイドライド(H^-)移動, aromatization 芳香族化, chain reaction 連鎖反応, initiation 開始, propagation 成長, termination 停止, chain transfer 連鎖移動, coupling カップリング, recombination 再結合, disproportionation 不均化

 ex. $RCH_2CH_2 \cdot + \cdot CH_2CH_2R \rightarrow RCH_2CH_2 - CH_2CH_2R$:ラジカルのカップリングまたは再結合
   $RCH_2CH_2 \cdot + \cdot CH_2CH_2R \rightarrow RCH_2CH_3 + CH_2 = CHR$:ラジカルの不均化

cracking 分解, decomposition 分解, degradation 分解, 崩壊, 退化

 cf. bio-degdarable plastics 生分解性プラスチック

polymerization 重合, polycondensation 重縮合, polyaddition 重付加
depolymerization 解重合, dimerization 2量化

 cf. monomer モノマー(単量体), dimer ダイマー(2量体), trimer トリマー(3量体), tetramer テトラマー(4量体)

in situ その場(系内)で, in vivo 生体内で, in vitro 生体外(試験管内)で, catalyze (*vt*)触媒(接触)する, catalysis (*n*)触媒(接触)反応, 触媒作用, catalyst (*n*)触媒, yield (*vt*)を生じる(産する), (*n*)収量, 収率, produce, give, afford (*vt*)産する, 与える, 生じる, give rise to~ ~を生じる, ~の原因となる, reactant 反応物, starting material 出発物質, feed 原料(仕込み), charge 仕込み(量), product (*n*)生成物, 製品, production 生産, conversion (*n*)収率, 変換率, convert (*vt*)変換する, rate of reaction 反応速度, rate constant 速度定数, rate-determining 律速の, equilibrium 平衡, equilibrium constant 平衡定数, activate (*vt*)活性化する, activation energy 活性化エネルギー, heat of reaction 反応熱, identification 同定, isolation 単離, purification 精製

一般用語・熟語

 account for~:~を説明する, ~の原因となる,

～due to ―：―に基づく～

be attributed to～：～に起因する．

be responsible for～：～の原因となる．

result in～：～の結果となる

available(a)入手できる，入手可能な，利用できる

apparent(見かけ上)明らかな，apparently(見かけ上)明らかに

be allowed to ―：― させる

 ex. The reaction mixture was allowed to stand overnight.　反応混合物は一夜放置させた。

cause～(to)―　：　～に(を)― させる

 ex. Virus molecules have the power of self-duplication, causing other molecules identical with themselves to be formed.
 ウイルス分子は自己複製力があり，自分と同じ他の分子を生成させる。

force～(to)―　：　(無理に)～に(を)― させる(強要)

 ex. When the rubber is stretched, the individual polymer chains are forced to change to a much more extended shape.
 ゴムを延伸すると，個々のポリマー鎖は，(無理やり)はるかにもっと引き伸ばされた形に変えられる。

certain 確かな，certainly 確かに，

a certain～：ある ～

 ex. a certain metal oxide ある金属酸化物，a certain Mr. Dewar デュワーさんという人

characteristic of～：～に特有の

 ex. Association due to hydrogen-bonding is responsible for the very high boiling points characteristic of hydroxylic compounds as compared with those of hydrocarbons with comparable molecular weights.
 水素結合に基づく会合が，同様の分子量を持った炭化水素と比べて，ヒドロキシ化合物特有の非常に高い沸点の原因である。

comparable(to～)：(～と)同じぐらいの，(～と)比較し得るほどの

 ex. The metallic bond in mercury is as weak as some intermolecular forces, while that in tungsten is comparable in strength to a multiple covalent bond.
 水銀の金属結合はいくつかの分子間力と同じぐらいに弱いが，一方，タングステンのそれ(金属結合)は多重結合と同じぐらいの強さである。

as～as：と同じほどの

considerable, appreciable：かなりの

consist of～：(*vi*)～から成る，　constitute～, comprise～：(*vt*)～を構成する

 ex. DNA molecules consist of sugar (deoxyribose), base (A, T, G, C) and phosphate linkage.
 DNA分子は，糖(デオキシリボース)，塩基(A, T, G, C)，リン酸エステル結合から成

る。
Sugar (ribose), base (A, U, G, C) and phosphate linkage constitute an RNA molecule.
糖(リボース),塩基(A, U, G, C),リン酸エステル結合がRNA分子を構成する。

provided (that) 〜：もし〜ならば，〜という条件で
    ex. The reaction goes, provided the temperature is sufficiently high. その反応は，温度が十分高ければ，うまくいく。

subject to 〜：〜を受けやすい
    ex. Double bonds are subject to oxidation. 二重結合は酸化を受けやすい。

A followed by B： Aに続いてB，Aのあとで B. ＝ B follow(s) A
    ex. The polymers were isolated by precipitation from benzene into methanol, followed by vacuum drying.
ポリマーは，ベンゼンからメタノールに沈殿の後，真空乾燥によって，単離された。

〜, while ―；〜, whereas ― ： 〜(文章)，一方―(文章)
    ex. Stronger acids such as hydrochloric acid react almost completely with water, whereas weaker acids such as acetic acid react only slightly.
塩化水素のような強酸は水と殆ど完全に反応するが，一方，酢酸のような弱酸はほんの少ししか反応しない。

3 よく使われる構文

not only 〜, but (also) ― ：〜ばかりでなく，―も
    ex. His research was important not only for its theory, but for its practical applications.
彼の研究は，その理論ばかりでなく，実用的応用でも重要であった。

too 〜 to ―：あまりに〜(*a* 形容詞 または *ad* 副詞)なので，―(*v* 動詞)できない。(―するには，〜過ぎる)
    ex. The boiling points of the three xylenes are too close to be separated by usual distillation.
3つのキシレンの沸点は近すぎるので，ふつうの蒸留で分けることはできない。

so 〜 that ―：あまりに〜(*a* 形容詞 または *ad* 副詞)なので，―(文章)である。
such 〜 that ―：あまりに〜(*a* 形容詞 または *ad* 副詞)なので，―(文章)である。
―(文章)であるほどに，〜(*a* 形容詞 または *ad* 副詞)である。
    ex. Their boiling points are so close that they cannot be separated by usual distillation.
They have such close boiling points that they cannot be separated by usual

distillation.
それらの沸点は近すぎるので，ふつうの蒸留で分けることはできない。

it is～(for -)to ― : (- が)―するのは～である。
　it は形式(仮)主語((for -)to ―の代名詞)

ex. It is difficult (for him) to complete the reaction in time to prepare the objective compound.
目的の化合物を間に合うように合成するために，(彼が)その反応を完成させるのは困難であった。

that のさまざまな使われ方
● 代名詞として：

ex. Without supercooling, the temperature at which a pure solid melts is identical with that at which its melt freezes.
過冷却がなければ，純粋な固体が融ける温度は，その融解物が凍るそれ(温度)と同じである。

● 関係代名詞として：

ex. Because of amide linkages, nylon is a synthetic polymer that is a biological analog.
アミド結合の故に，ナイロンは生物類似体の合成高分子である。

● 接続詞として：主語，目的語，補語，同格の節を率いる。

　　it appears that～, it seems that～ : ～(文章)と思われる，～(文章)らしい。
　　it is thought that～, it is considered that～ : ～(文章)と考えられる。
　　it is certain that～, it is not certain that～ : ～(文章)は確かである。確かでない。
　　　(上記3つの例で，it は that 以後の仮主語)

ex. It appeared that a certain mixture of metal oxides increased the rate of reaction enough for the process to be put into practice.　ある種の金属酸化物混合物がその反応速度を十分に増加させ，その工程を実施できるほどにする と思われた。
The dogma was that the order of the genes is immutable.
ドグマ(定説)は，遺伝子の配列は不変であるということであった。
The increased rate indicated that the catalyst worked well. その速度増加は，その触媒がうまく働いた事を示す。
The catalyst was so designed that the decomposition was enhanced enough for commercial use.
The catalyst was designed in order that the decomposition was enhanced enough for commercial use.
その分解が商業的に十分利用できるほどに加速されるように(目的で)，その触媒は設計された。

cf. The catalyst was designed in order to enhance the decomposition enough for commercial use.
工業的に十分利用できるほどにその分解を促進するように(目的で)，その触媒は設計された。

英文法を説明すると切りがないが，ごく簡単に復習すると，

不定詞（to + 動詞）

名詞的用法：

主語として　*ex.* **To** complete the reaction is difficult for him.
反応を完結させるのは，彼には難しかった。
（上記の it is〜(for -)to ─ it も参照）。

目的語として　*ex.* He tried **to** complete the reaction.
彼はその反応を完結させようとした。

形容詞的用法：*ex.* He tried the reaction **to** go well.
彼はその反応がうまくいくように試みた。

副詞的用法（目的，結果を意味する場合が多い）：

ex. The reaction was tried **to** prepare the objective compound.
その反応は，目的化合物を合成するために試みられた。
The reaction went well **to** prepare the objective compound.
その反応はうまくいって，目的化合物が合成できた。

動名詞（動詞 -ing）：〜すること（動詞を名詞として使う）

ex. **Raising** the temperature has made the reaction proceed completely. 温度を上げることが，その反応を完全に進行させた。
注：次の例文中 raising も参照。また，proceed は原型不定詞で，to が省略されている。

現在分詞（動詞 -ing）：状態，理由，結果などを意味する付随的用法が多い。

ex. By **raising** the temperture, he confirmed the reaction **going** well.
温度を上げることで，彼はその反応がうまくいくのを確かめた。

進行形（be + 動詞 -ing）：今まさに〜している，〜していた，という現在，過去の進行状況に使われる。

ex. The reaction **is going** to start.　反応は今まさに始まろうとしている。
The reaction **was going** to be completed.　反応は完了しようとしていた。

受動態（受身）（be + 過去分詞）：とくに主語不特定（一般的な人や物）の場合に，この形がよく使われる。

過去分詞：先行あるいは後続する名詞を修飾する形容詞的利用が多い。

ex. The **emitted** light was **absorbed** by the substance present there.
発せられた光はそこにある物質に吸収された。
The light **absorbed** made the substance brittle.
吸収された光はその物質をもろくした。

不定詞，動名詞，現在分詞，受動態は，化学論文においても頻繁に使われる。次の例文のアンダーラインを付した青字の部分をいずれの場合であるかチェックしてみよう。この短文に，不定詞が3回，動名詞5回，過去分詞を含めた受動態9回が現れる。

[Ex.1][1)] Identifying Molecules and Atoms with Light

The specific wavelengths of light absorbed or emitted by a particular molecule or atom are unique to that molecule or atom and can be used to identify it. By extending the range of light used into the UV, infrared, and radio wave regions, virtually any element or compound can be identified by its interaction with light. Scientists use the interaction of light with matter, called spectroscopy, to identify unknown substances.

Astronomers, for example, identify the elements present in faraway stars by examining the light emitted by the star. Chemists can determine the composition of a particular substance by studying the wavelengths of light absorbed by the substance. Atmospheric scientists study slight variations in the colors present in sunlight to identify the molecules present in the atmosphere. Spectroscopy is one of the most versatile tools a scientist has at his or her disposal for identifying and quantifying matter.

[注]　（青色の単語および熟語の和訳例）
specific 特有の，明確な，absorb 吸収する，emit 発する，particular 特定の，
identify 同定する，extend 拡大する，virtually 実質的に，interaction 相互作用，
spectroscopy 分光学，to identify（不定詞，目的または結果）同定するために，
astronomer 天文学者，composition 組成，atmosphere 大気，versatile 多芸多才な，
at one's disposal の自由に（使える），quantify 定量する，
extending, examining, studying, identifying, quantifying：（いずれも動名詞）　～すること
present：存在する（現在分詞 being が省略されている：先行する名詞（molecules）を修飾）

和訳例

光を用いる分子および原子の同定

　ある特定の分子あるいは原子によって吸収され，または発せられた光の特有の波長は，その分子あるいは原子に独特のものであり，それを同定するのに用いることができる。用いる光の範囲を UV（ultraviolet 紫外），赤外，そしてラジオ波領域に拡大することによって，実質的にどんな元素あるいは化合物も，それと光との相互作用によって同定することができる。科学者は，分光学と呼ばれる，光と物質の相互作用を利用して，未知物質を同定する。

　たとえば，天文学者は，遠くの星にある元素を，その星の発する光を調べることによって同定する。化学者は，ある特定の物質の組成を，その物質の吸収する光の波長を研究して決定することができる。大気の科学者は，日光に存在する色のわずかな変化を研究して，大気中にある分子を同定する。分光学は，物質を同定し，定量するため，科学者が自由に使える最も多才な道具の一つである。

4 基本英文型

5つの基本文型がある。主語(S = Subject)，動詞(V = Verb：Vi = 自動詞，Vt = 他動詞)，補語(C = Complement)，目的語(O = Object)，を明らかにしてみよう。

(例文中，アンダーラインした単語が対応するS, V, C, O)

第1文型：S + Vi　　　ex.　Petroleum occurs in nature.　石油は天然に産出する。
第2文型：S + Vi + C　ex.　It is a mixture of hydrocarbons.　それは，炭化水素の混合物である。
第3文型：S + Vt + O　ex.　It has produced many petrochemicals.　それは多くの石油化学製品を生み出してきた。
第4文型：S + Vt + O + O
　　　　　　　　　　　ex.　It brought us prosperity.　それは私たちに繁栄をもたらした。
第5文型：S + Vt + O + C
　　　　　　　　　　　ex.　Petroleum has made us happy.　石油は私たちをハッピーにしてきた。

そして，それらにつく品詞(冠詞，名詞，形容詞，副詞)，および修飾する句，節，挿入句，挿入節，同格語句・節などを明らかにすることも，込み入った文章や，長文の解釈の助けとなる。上例に関連して，次の文(一部，文献[2]から)があったとしよう。アンダーライン以外は，冠詞，形容詞，副詞，修飾句・節である。

[Ex. 2]　Petroleum

Petroleum so far has occurred in abundance in nature. It is a mixture of a number of hydrocarbons. It has produced many kinds of petrochemicals that are useful in our modern life. In this way, it has brought us prosperity and made us happy. However, we are now in a terrible dilemma due to acute oil shortages worldwide. A petroleum substitute is urgently needed and one of promising candidates is nuclear energy

［解説］

Petroleum so far has occurred in abundance in nature. It is a mixture of a number of hydrocarbons.
　S　　　　　　Vi(現在完了)　　　　　　　　　　　　S Vi　　C
　石油は　今まで　産出してきた　多量に，天然に　　　　それは混合物である　　数多くの炭化水素の
It has produced many kinds of petrochemicals that are useful in our modern life.
S　Vt(現在完了)　　　　　O　　　　　　　　　　(S) (Vi)　(C)
それは生産してきた　石油化学製品の多くの種類を　　我々の現代生活に有用な
　　　　　(that は関係代名詞で接続詞，petrochemicals の代名詞，主語の役割をしている。)

```
In this way, it has brought   us prosperity and made us happy.
              S   Vt(現在完了)  O  O           Vt   O  C
  このように  それはもたらしてきた   我々に繁栄を  そして  我々をハッピーにした
However, we are now in a terrible dilemma due to acute oil shortages worldwide. A petroleum substitute
         S  Vi        C                                                        S
  しかし  我々は 今  恐ろしいジレンマの中にいる   世界中の深刻な石油不足のために    石油代替品が
is urgently needed and one of promising candidates is nuclear energy
      Vi (Vtの受身)     S                        Vi    C
   緊急に必要とされている   そして  有望な候補の1つは  核エネルギーである
```

通して，和訳すると，次のようになる。

和訳例

石　油

　　石油は今まで天然に多量に産出してきた。それは，数多くの炭化水素の混合物である。それは，私たちの現代生活に有用な多種の石油化学製品を生み出してきた。このように 石油は私たちに繁栄をもたらし，私たちをハッピーにしてきた。しかし，今私たちは，世界中の深刻な石油不足のために，恐ろしいジレンマ(板ばさみの中)にある。石油代替品が緊急に必要とされ，一つの有望な候補は核エネルギーである。

　以下，化学に関係する基本的な英文のテキストから，さまざまな場面を引用して，例題(Exercise)とする。読者は，まず，英文を読み(できれば，声に出して読むことも重要である，本書では発音に関してはまったく省略したが)，大筋をつかんだ上で，自らの和訳を試みよう。英文のあとには，鍵となる単語，熟語，構文の注をつけ，さらに，この基礎編の例題では，ところどころ，上記の基本文型を含めた解説を付した。例題順にこだわらず，内容・興味に応じて，先へ進むことも，英文に慣れるための良法である。

例題の凡例

　［注］は，初心者には馴染みの少ない，または難解と思われる単語・熟語，また重要な構文について，その英文に最適と思われる注訳，または最も標準的と思われる注釈を示した。ただし，これらは使われている情況によって，意味の変わる場合もあるので，自身で辞書を引いて，最適の訳語を把握する習慣をつけることが重要である。また，同じ［注］を何度も複数の例題に繰り返した場合もあることをお断りしておきたい。
　［解説］には，基本文型のSVOC，単語・区・節ごとの和訳例などを示した。 ↰ は先行語句を修飾することを表し，S, V, O, Cは主文中，(S)，(V)，(O)，(C)は修飾節または挿入節中のSVOCの対応を示す。
　［和訳例］は，例題ごとの通訳を示した。

[Ex. 3]　[3)]General Chemistry

偉大な先達，Linus Pauling (1901-1994；1954ノーベル化学賞，1962ノーベル平和賞

をそれぞれ単独で受賞）の著名な"General Chemistry"（1970年版）[3]と，化学と直接の関係はないが，この偉人の子供のころを著した伝記[4]から引用して，基礎編への導入としよう．

1. Matter and Chemistry

The universe is composed of matter and radiant energy. Matter (from the Latin *materia*, meaning wood or other matter) may be defined as any kind of mass-energy that moves with velocities less than the velocity of light, and radiant energy as any kind of mass-energy that moves with the velocity of light.

The different kinds of matter are called substances. Chemistry is the science of substances — their structure, their properties, and the reactions that change them into other substances.

This definition of chemistry is both too narrow and too broad. It is too narrow because the chemist in his study of substances must also study radiant energy, in its interaction with substances. He may be interested in the color of substances, which is produced by the absorption of light. Or he may be interested in the atomic structure of substances, as determined by the diffraction of X-rays or by the absorption or emission of radiowaves by the substances.

On the other hand, the definition is too broad, in that almost all of science could be included within it. The astrophysicist is interested in the substances that are present in stars and other celestial bodies, or that are distributed, in very low concentration, through interstellar space. The nuclear physicist studies the substances that constitute the nuclei of atoms. The biologist is interested in the substances that are present in living organisms. The geologist is interested in the substances, called minerals, that make up the earth. It is hard to draw a line between chemistry and other sciences.

> ［注］　be composed of から成る，matter ［もの］，radiant energy 放射エネルギー，mass-energy 質量―エネルギー，substance 物質，diffraction 回折，absorption 吸収，emission 放射，astrophysicist 天体物理学者，celestial 天の，interstellar 星間の，constitute 構成する

［解説］

Paulingの物質・宇宙感を彷彿とさせる文章である．matterとsubstanceは，いずれも通常「物質」と訳されるが，Paulingは，3で見るように，matterをsubstanceよりも広い意味で使っており，ここでは「もの」と訳した．また，mass-energyは，2で見るように，アインシュタインの相対性理論にある質量―エネルギー変換則に基づいた普遍的定義からきている．4では，科学と社会・政治の問題に触れている．今，どこかの国の，いや世界の指導者にも聞かせたい慧眼ではないだろうか．

関係代名詞，接続詞などを含む複文が多いが，各文（節）は単純な文型の構成から成る．2文のみを挙げて，解説しよう．

Linus Carl Pauling
http://nobelprize.org/nobel_prizes/peace/laureates/1962/pauling.html

Matter	may be defined	as any kind of mass-energy	that moves with velocities less than the velocity of light,
「もの」は	定義されるだろう	あらゆる種類の質量―エネルギーとして	◁光の速度より小さい速度で動く
S	V(Vtの受動態)	C	thatは関係代名詞で*conj*と(S)を兼ねる　movesが(Vi)の第1文型の節

and	radiant energy		as any kind of mass-energy	that moves with the velocity of light.
そして	放射エネルギーは		あらゆる種類の質量―エネルギーとして	◁光の速度で動く
conj.	S	間に may be defined が省略されている	C	前文節と同様の接続節

> [!NOTE] 和訳例

「もの」と化学

　宇宙は，「もの」と放射エネルギーから成る。「もの」（木や他のものを意味するラテン語 *materia* から）は，光速よりも小さい速度で動く，あらゆる種類の「質量―エネルギー」と定義され，放射エネルギーは光速で動く，あらゆる種類の「質量―エネルギー」と定義されるだろう。

　さまざまな種類の「もの」は物質と呼ばれる。化学は物質―その構造，その性質，そしてそれらを別の物質に変える反応―の科学である。

　この化学の定義は，あまりに狭すぎ，またあまりに広すぎる。あまりに狭いというのは，物質の研究をしている化学者は，その物質との相互作用という点で，放射エネルギーも研究しなければならないからである。彼は，光の吸収によって生じる物質の色に興味があるかもしれない。あるいは，彼は物質の原子構造に興味をもつこともあるだろう。これはX線回折や，その物質による電波の吸収や放射によって決定される。

　一方で，ほとんどすべての科学がその中に含まれ得るという点で，その定義はあまりに広すぎる。天体物理学者は，星や他の天体に存在する物質，あるいはまた星間の空間を通して非常に低濃度で分布する物質に興味がある。核物理学者は，原子の核を構成する物質を研究する。生物学者は，生物有機体に存在する物質に興味がある。地理学者は，地球を形作る鉱物と呼ばれる物質に興味がある。化学とほかの科学の間に線を引くのは難しいのである。

2. Mass and Energy

The amount of mass associated with a definite amount of energy is given by an important equation, the *Einstein equation*, which is an essential part of the theory of relativity :　　　$E = mc^2$

In this equation E is the amount of energy (J), m is the mass (kg), and c is the velocity of light (m s^{-1}). The velocity of light, c, is one of the fundamental constants of nature; its value is 2.9979×10^8 meters per second.

Until the present century it was thought that matter could not be created or destroyed, but could only be converted from one form into another. In recent years it has, however, been found possible to convert matter into radiant energy, and to convert radiant energy into matter. The mass m of the matter obtained by the conversion of an amount E of radiant energy or convertible into this amount of radiant energy is given by the Einstein equation. Experimental verification of the Einstein equation has been obtained by the study of processes involving nuclei of atoms.

　［注］　associated with に関連した，definite 一定の，theory of relativity 相対性理論，
　　　　be converted from ― into ～：―から～に変換される，convert ― into ～：―を～に変

換する，conversion 変換，verification 証明

> 和訳例

質量とエネルギー
　ある一定のエネルギー量と関連した質量の量は，重要な式，アインシュタイン式で与えられる。これは相対性理論の本質の部分である：$E = mc^2$。この式で，E はエネルギー量(J)，m は質量(kg)，c は光の速度($m\ s^{-1}$)である。光速 c は基本的な自然定数であり，その値は１秒当たり 2.9979×10^8 メートルである。
　今世紀(注：20世紀)までは，「もの」は創造されたり，破壊されたりすることはできず，１つの形から別の形へ変えることだけができる，と考えられていた。しかし，近年になって，「もの」を放射エネルギーに換え，放射エネルギーを「もの」に換えることが可能であることが見つけられた。放射エネルギー量 E の変換で得られるか，あるいは，この放射エネルギー量に変換できる「もの」の質量 m は，アインシュタイン式で与えられる。アインシュタイン式の実験的証明は，原子核を含む過程の研究によって得られた。

3 Matter and Substance, and Kinds of Definition

We shall first distinguish between objects and kinds of matter. An object, such as a human being, a table, a brass doorknob, may be made of one kind of matter or different kinds of matter. The chemist is primarily interested not in the objects themselves, but in the kinds of matter of which they are composed. He is interested in the alloy brass, whether it is in a doorknob or in some other object; and his interest may be primarily in these properties of the material that are independent of the nature of the objects containing it.

A substance is usually defined by chemists as a homogeneous species of matter and reasonably of definite chemical composition.

Definition may be either precise or imprecise. The mathematician may define the words that he uses precisely ; in his further discussion he then adheres rigorously to the defined meaning of each word. One of precise definitions in chemistry is that of the kilogram in SI system as the mass of a standard object, the prototype kilogram, that is kept in Paris. The gram is defined rigorously and precisely as 1/1000 the mass of the kilogram.

On the other hand, the words that are used in describing nature, which is itself complex, may not be capable of precise definition. In giving a definition for such a word the effort is made to describe the accepted usage.

　　［注］　distinguish between A and B：A と B を区別する，object 物体，対象，human being 人間，brass 真ちゅう，be interested in に興味がある，primarily 主に，not — but 〜：—でなく，〜，be composed of から構成される，alloy 合金，whether — or 〜：—であれ，〜であれ，homogeneous 均一な，adhere to に固執する，rigorously 厳密に，prototype 原型，complex 複雑な

和訳例

「もの」と物質，そして定義の種類．

　我々は，まず物体と「もの」の種類を区別することにする．人間，テーブル，真ちゅうのドアノブといった物体は1種類の「もの」あるいはさまざまな種類の「もの」からできている．化学者は主に，物体そのものに興味はなくて，それらが構成されている「もの」の種類に興味がある．彼は，ドアノブであれ，ある他の物体であれ，その中にある合金の真ちゅうに興味があるのである．そして，彼の興味は主に，それを含む物体の性質には無関係な材料の性質にあるのである．

　物質とは，通常化学者によって均一な「もの」で，合理的に一定の組成から成る化学種として定義されている．

　定義は，正確か不正確かどちらかである．数学者は，使う言葉を正確に定義する；議論を進めるさいに，彼は，各言葉の定義された意味に強く固執する．化学における正確な定義の1つは，SI 単位系のキログラム（注：Ex. 26 参照）で，標準物体すなわちパリに保存されている標準物体，キログラム原型（注：白金—イリジウム合金でできている）の質量としての定義である．グラムは，キログラム質量の 1/1000 として厳密にそして正確に定義される．

　一方，それ自体が複雑な自然を表すのに用いられる言葉は，正確な定義ができないかもしれない．このような言葉の定義を与えるさいには，〈一般に〉認められた用法を言葉にする努力がなされる．

4. Science and Society

The application of the scientific method does not consist solely of the routine use of logical rules and procedures. Often a generalization that encompasses many facts has escaped notice until a scientist with unusual insight has discovered it. Intuition and imagination play an important part in the scientific method.

As more and more people gain a sound understanding of the nature of the scientific method and learn to apply it in the solution of the problems of everyday life, we may hope for an improvement in the social, political and international affairs of the world. Technical progress represents one way in which the world can be improved through science. Another way is through the social progress that results from application of the scientific method — through the development of "moral science." I believe that the study of science, the learning of the scientific method by all people, will ultimately help the people of the world in the solution of our great social and political problems.

［注］ consist of から成る，routine おきまりの，logical 論理的な，generalization 一般化，encompass 包含する，escape 逃れる，intuition 直感，improvement 改善，affair 事柄

和訳例

科学と社会

　科学的方法の応用は，単に，論理的な規則や手法のおきまりの利用のみから成るものではない．しばしば，多くの事実を包含している一般則が，非常な洞察力をもった科学者が発見するまで注目を浴びなかったことがある．直感と想像は，科学的方法の重要な役割を演じる．

より多くの人々が科学的方法の性質について健全な理解を得，日常の生活の問題解決にそれを応用することを学ぶにつれて，世界における社会的，政治的，そして国際間のできごとに改善を望むであろう。技術の進歩は，世界が科学を通して改善され得る1つの道を表している。もう1つの道は，科学的方法の応用からくる社会の進歩—「倫理の科学」の広がりを通してである。私は信じる，科学の研究，すべての人々による科学的方法の学習は，究極的には世界の人々を，現在の大きな社会的，政治的問題の解決という点で，手助けするであろうことを。

[Ex. 4][4) The Early Days of Linus Pauling

1. Linus Carl Pauling is a quintessentially American figure. Born into conflict and combat with the rawest elements of nature, poverty, and disease, he toughened himself for a lifetime of struggle. Brilliant, undisciplined, and rebellious, Pauling burst quickly onto the scientific scene. Unflappable in the face of criticism, unshakably confident in his own genius, Pauling challenged scientific orthodoxy armed with a minimum of hard scientific data and a tidal surge of intellectual power. Before reaching the age of thirty, he ranked among the great scientists of the world.

　　［注］　quintessentially 生粋に，poverty 貧乏，undisciplined 規律にしばられない，rebellious 反体制の，unflappable 冷静な，unshakably ゆるぎなく，orthodoxy 伝統，armed with で武装した，tidal 潮の，surge 大波

[和訳例]

ライナス・ポーリングの幼少の日々．

　ライナス・カール・ポーリングは生粋のアメリカ人である。自然の中で最も不当な要素すなわち貧乏と病気との闘争と格闘の中に生まれ，彼は，闘いの人生を送るための強さを自ずと養っていた。聡明で，現状に飽き足らず，妥協しないで，ポーリングは早くも科学分野で頭角をあらわした。批判に直面しても冷静で，自分の資質にゆるぎなく確信をもつポーリングは，最少の確かな科学的データと大波のような知力で武装して，科学の伝統に挑戦した。30歳になる以前に，彼は世界の偉大な科学者の中に位置したのである。

2. If one wishes to speculate about the often brilliant, always quixotic, and occasionally stormy career of Linus Pauling, his maternal relatives provide some fascinating clues. They were an odd bunch, as bizarre as the paternal side was humdrum. There was old Will Darling, Linus Darling's brother, who was a painter and a confirmed spiritualist. Always "in touch" with a dead Indian named Red Cloud, he hoped to find the location of a lost gold mine in the Lone Rock country. Not to be outdone, Linus Pauling's aunt, Stella "Fingers" Darling was known throughout the West for her ability to break open safes. If Linus Pauling's later actions in politics and science were fabulous and unorthodox, he was not, at least, breaking with family tradition.

　　［注］　quixotic ドンキホーテ流の（衝動的な），maternal 母方の，paternal 父方の，odd 風変わりな，bunch 集まり，bizarre 奇妙な，humdrum 平凡な，confirmed 頑固な，

spiritualist 降霊術者，mine 鉱山，not to be outdone 負けることない，safe 金庫，fabulous 伝説的な，unorthodox 非正統的な，break with 断絶する

和訳例

　しばしば輝かしく，常にドンキホーテ流で（衝動的で），時に嵐のような，ライナス・ポーリングの生涯について推測したいなら，彼の母方（ダーリン家）の親戚がいくつかのうなずけるような手がかりを与えてくれる。彼（彼女）らは，父方（ポーリング家）が平凡であったと同じぐらい奇妙に，風変わりな集まりであった。ライナス・ダーリン（注：ライナスポーリングの母の父）の兄（弟）である年老いたウィル・ダーリンがいた。彼は絵描きで，頑固な降霊術者であった。いつもレッド・クラウドという名前の死んだインディアンに「触り」ながら，ローン・ロック地方の失われた金鉱山を見つけようとしていた。それに負けずに，ライナス・ポーリングのおば，ステラ・"フィンガー（指）"・ダーリンは，金庫を開けるという（指の）能力で西部じゅうに知られていた。もしライナス・ポーリングの政治・科学における後の行動が伝説的で，非正統的であるとすれば，彼は，家族の伝統と，少なくとも，断絶したわけではない。

3. Pauling denies any very early interest in chemistry. In an interview given many years later, he said that he was about twelve years old when he became interested in chemistry. His father died when he was nine. On the other hand, he does admit to an early interest in science generally. In the interview, he says: "My interest in mathematics was already evident and of course I found that scientific questions interested me as a boy. I was always puzzling about various phenomena such as shadows, optical phenomena that I had observed of one sort or another." Perhaps some sort of "imprinting" took place very early in Pauling's life as he watched his father mix medicine, though he had no interest in chemistry until later.

　［注］　inprinting 刷り込み，take place 起こる，one…another あれやこれやの

和訳例

　ポーリングは非常に早くから化学への興味を持ったことを否定している。何年もあとのインタビューで，およそ12歳になって，化学に興味を持つようになったと言っている。彼の父は，彼が9歳のときに亡くなった。一方で，彼は科学一般には早くからの興味を認めている。インタビューで彼の言うには，「数学への私の興味はすでに明らかで，もちろん，科学的な疑問が少年の私を惹きつけたことを知っていた。私はいつも影，あれやこれやで私が見た光学的現象，のようなさまざまな現象を不思議に思っていた」。多分，父が薬を混ぜるのを見て，ポーリングの生活の早い時期に，ある種の「刷り込み」がなされたのであろう。もっとも，もっと後になるまで化学には何の興味も持たなかったけれども。

4. Interestingly, there is evidence that Pauling felt some attraction to religion in his early years. In the same interview he says: "I can remember lying in bed in my grandmother's house looking at a …drawing of a head of Christ…here I was looking at this head of Christ when I saw that a halo appeared above it, a band of light…after …investigation I discovered the after-image effects…of the retina and satisfied myself that this was a general natural phenomenon…" Evidently, whatever environmental pressure there was to practice religion soon resolved itself, and his

humanistic and anti-metaphysical tendencies appeared soon after, when he was still a boy.

［注］　halo かさ，後光，after-image 残像，retina 網膜，metaphysical 形而上学の

和訳例

　興味深いことに，ポーリングは早い時期に宗教に対してある魅力を感じたという証拠がある。同じインタビューで，彼は言う：「私は，キリストの頭部の絵を見ながら，祖母の家のベッドに寝そべっているのを思い出す。キリストの頭部を見ていたときに，その上に後光が現れるのを見た。光の帯…後で…調べて，残像効果であることを発見した…網膜の，そしてこれは一般的な自然現象であると納得した…」。明らかに，宗教を慣習にする何らかの環境の圧力があったことが間もなく自明になった，そしてすぐ後で彼のヒューマニズム的，反形而上学的な傾向は，まだ少年であったときに，芽生えた。

5. Unlike the stereotypical scientist who inhabits a solipsistic world of his own creation, Linus Pauling was active, sports-oriented, and enjoyed the outdoors. On the rare sunny Oregon afternoon, he and his cousin Mervyn Stephenson would wander aimlessly through the frontier streets of Condon, swiping at the occasional butterfly and skipping rocks over the waters of nearby ponds. In the summers, the pair spent a great deal of time on the wheat ranch of Mervyn's father, Philip Hebert Stephenson. Life at the ranch was carefree, punctuated by occasional hunting expeditions and swims in the nearby streams.

［注］　stereotypical 紋切り型の，inhabit に住む，solipsistic 唯我論の，swipe at を大振りに打つ，盗む，ranch 農場，punctuate 中断する，expedition 旅行

和訳例

　自分の創った唯我論の世界に住み慣れた紋切り型の科学者とは違って，ライナス・ポーリングは活発，スポーツ志向であり，アウトドアを楽しんだ。まれにある日当たりの良いオレゴンの午後には，彼と彼のいとこのメルビン・ステフェンソンは，当てもなく，コンドンの辺境の通りをぶらつき，時々の蝶々を追いかけ，近くの池の水の上の岩を飛んだりしていた。夏には，二人は多くの時間を，メルビンの父，フィリップ・ヘバート・ステフェンソンの小麦の農場で過ごした。農場での生活は気楽で，合間には時々の狩りの旅行や近くの川での水泳を楽しんだ。

[Ex.5][5)]　Nagasaki 1945

I discovered the green fluorescent protein GFP from the jellyfish *Aequorea aequorea* in 1961 as a byproduct of the Ca-sensitive photoprotein acquorin, and identified its chromophore in 1979. GFP was a beautiful protein but it remained useless for the next 30 years after the discovery.

My story begins in 1945, the year the city of Nagasaki was destroyed by an atomic bomb and World War II ended. At that time I was a 16-year old high school student, and I was working at a factory about 15 km northeast of Nagasaki. I watched the B-29 that carried the atomic bomb heading toward Nagasaki, then soon I was exposed to a blinding bright flash and a strong pressure wave that were caused by a gigantic explosion. I was lucky to survive the war. In the mess after the war, however, I could not find any school to attend. I idled for 2 years, and then I learned that the pharmacy school of Nagasaki Medical College, which had been completely destroyed by the atomic bomb, was going to open a temporary campus near my home. I applied to the pharmacy school and was accepted. Although I didn't have any interest in pharmacy, it was the only way that I could have some education.

After graduation from the pharmacy school, I worked as a teaching assistant at the same school, which was recognized as a part of Nagasaki University. My boss Professor Shungo Yasunaga was a gentle and very kind person. In 1955, when I had worked for four years on the job, he arranged for me a paid leave of absence for one year, and he sent me to Nagoya Univerity, to study at the laboratory of Professor Yoshimasa Hirata. The research subject that professor Hirata gave me was the bioluminescence of the crustacean ostracod *Cyypridina hilgendorfii*.

　［注］　the pharmacy school of Nagasaki Medical College　長崎医科大学薬学専門部，temporary 仮の，bioluminescence 生物発光，crustacean osttrcod 甲殻類貝虫（カイムシ）

［解説］下村修氏のノーベル賞講演（2008）"Discovery of Green Fluorescent Protein, GFP（緑色蛍光タンパク質，GFP，の発見）"の冒頭からである．一部，対訳を示す．

I discovered the green fluorescent protein GFP from the jellyfish *Aequorea aequorea* in 1961　as a byproduct of the
私は 1961 年くらげ（*Aequorea aequorea* エクオレア）からの緑色蛍光タンパク質（GFP）を発見した　　　カルシウム感受性の
Ca-sensitive photoprotein acquorin, and identified its chromophore in 1979. GFP was a beautiful protein
光たんぱく質エクオリンの副生物として　そしてその発色団を 1979 年に同定した．　GFP は美しいタンパク質である
but it remained useless for the next 30 years after the discovery.
が，しかし，発見後の 30 年間は　無用のままであった．
　　　My story begins in 1945, the year the city of Nagasaki was destroyed by an atomic bomb and World War II ended. At the
　　　私の話は 1945 年に始まる．すなわち，長崎市が原子爆弾で破壊され，第 2 次世界大戦が終わった年である．　その
time I was a 16-year old high school student, and I was working at a factory about 15 km northeast of Nagasaki.
当時　私は 16 才の高校生でした　　　そして私は，長崎の北東およそ 15 km のある工場で働いていました．
I watched the B-29 that carried the atomic bomb heading toward Nagasaki,
私は，長崎の方向に原子爆弾を運んでいた B-29 を見ました
then soon I was exposed to a blinding bright flash and a strong pressure wave
それからすぐ，目もくらむ閃光と圧力波にさらされました．
that were caused by a gigantic explosion. I was lucky to survive the war. In the mess after the war, however,
それは巨大な爆発で引き起こされたものです．　私は幸運にも戦争を生き延びました．　しかし，戦後の混乱の中で，
I could not find any school to attend. I idled for 2 years, and then I learned that the pharmacy school of Nagasaki
私が通う学校はありませんでした．　　2 年間ぶらぶらしたあとで，私は知りました．　　長崎医科大学の薬学専門部が
College, which had been completely destroyed by the atomic bomb, was going to open a temporary campus near
　　　　　　　　原子爆弾で完全に破壊されていた　　　　　　　家の近くに仮のキャンパスを開こうとしていたことを

my home. I applied to the pharmacy school and was accepted. Although I didn't have any interest in pharmacy,
　　　　　　私はその薬学専門部を受験して合格しました。　　　　　　私は薬学に何の興味ももっていませんでしたが，
it was the only way that I could have some education.
それが　私が教育を受けることの出来る唯一つの道だったのです。

> [和訳例]

長崎 1945

　私は 1961 年くらげ (*Aequorea aequorea* エクオレア) からの緑色蛍光タンパク質 (GFP) を，カルシウム感受性の光タンパク質エクオリンの副生物として発見した。そしてその発色団を 1979 年に同定した。GFP は美しいタンパク質であるが，しかし，発見後の 30 年間は無用のままであった。

　私の話は 1945 年に始まる。すなわち，長崎市が原子爆弾で破壊され，第二次世界大戦が終わった年である。その当時，私は 16 才の高校生でした。そして私は，長崎の北東およそ 15 km の ある工場で働いていました。私は，長崎の方向に原子爆弾を運んでいた B-29 を見ました，それからすぐ，目もくらむ閃光と圧力波にさらされました。それは巨大な爆発で引き起こされたのです。私は幸運にも戦争を生き延びました。しかし，戦後の混乱の中で，私が通う学校はありませんでした。2 年間ぶらぶらしたあとで，原子爆弾で完全に破壊されていた 長崎医科大学の薬学専門部が 家の近くに仮のキャンパスを開こうとしていたことを，私は知りました。私は薬学専門部を受験して合格しました。私は薬学に何の興味も持っていませんでしたが，それが 私が教育を受けることの出来る唯一つの道だったのです。

　薬学専門部を卒業後，同じところで教育助手として働きました。専門部はその後，長崎大学の一部として認められました。私のボスの安永峻五教授は紳士的で親切な人でした。1955 年仕事を 4 年続けたとき，教授は 1 年間の給与付出張をアレンジして，名古屋大学の平田義正教授の実験室で研究するように送り出してくれました。平田教授が私に与えた主題は，甲殻類貝虫，海ボタル *Cyypridina hilgendorfii* の生物発光でした。

[Ex. 6]²⁾ Host-Guest Chemistry

　Donald J. Cram, Charles J. Pedersen, and Jean-Marie Lehn, working independently, shared the 1987 Nobel Prize in Chemistry for their work on "host-guest" chemistry.

　"The basis of our work", explains Lehn, "is the way molecules are able to recognize each other." In nature, molecules that work together have complementary shapes, like a lock and key, and only the right key will fit to initiate a given reaction.

　It seemed downright preposterous to Donald O. Cram, when he got a phone call notifying him that he had just won the Nobel Prize for Chemistry. Reason: Cram is in the carpet cleaning business. The Swedish Academy of Sciences had rung up the wrong man. Quipped UCLA chemist Donald J. Cram after hearing about the mix-up: "There is some chemistry involved in carpet cleaning."

　　[注] share 分け合う，recognize 認識する，each other お互いを，complementary 相補的な，相補う，downright まったくの，preposterous 途方のない，initiate 始める，quip 皮肉る，mix-up 混乱

　　[解説] もちろん，慣れれば，1 つ 1 つ S, V, O, C, 品詞などを確認する必要はなく，頭から順に，句・節ごとに対応して，内容を把握していくことが，論文解釈はもとより，講演，会

話に早く対応できる方法である。文法は，文章を論理的に解釈するための手段と考えて，要は内容の的確な把握が大事である。

　　鍵は，多くの英語・論文に接すること！　Practice makes perfect!

　　前半のみ，SVOC の解説をすると：

<u>Donald J. Cram, Charles J. Pedersen, and Jean-Marie Lehn</u>,　<u>working independently</u>,　<u>shared</u>　<u>the 1987 Nobel Prize</u>
　　　　　　　　　　　S　　　　　　　　　　　　　　　　　　（挿入句）　　　　　　　Vt　　　　　O
<u>in Chemistry</u>　　<u>for their work in "host-guest" chemistry</u>. <u>"The basis of our work"</u>,　<u>explains</u>　<u>Lehn</u>, "<u>is</u>　<u>the way</u>
　　　　　　　　　　　　Prize の修飾　　　　　　　　　　　　　　　S　　　　　　　　（Vt）　　（S）　　Vi　　C
　　（挿入句）
<u>molecules are able to recognize each other</u>."　　　　　　<u>In nature</u>, <u>molecules</u>　<u>that</u>　<u>work together</u>
　　(S)　　　(Vi)(C)　　　　　　　　　　　　　　　　　　　　　　　　　　S　　（conj. S）（Vi）
（way と molecules の間に関係代名詞が省略されている）　　　（molecules を修飾）
<u>have</u> complementary <u>shapes</u>, <u>like a lock and key</u>, and　only the right <u>key</u>　will　<u>fit</u>　to initiate a given reaction.
Vt　　　　　　　　　　O　　　　（挿入句）　　　　　　　　　　　　　S　　　　　　Vi

> [!NOTE] 和訳例

ホスト-ゲスト化学

　　Donald J. Cram, Charles J. Pedersen, および Jean-Marie Lehn は独立に研究していて，彼らの研究"ホスト-ゲスト"化学に対して，1987年ノーベル化学賞を分かち合った。Lehn の言うには，"私たちの研究の基礎は，分子がお互いを認識できる方法にある" と。自然界では，一緒に仕事をする分子は，鍵穴と鍵のように，相補的な形を持っていて，正しい鍵だけが与えられた反応を始めるのに適合するであろう。

　　Donald O. Cram にとってまったく途方もなく思えたのは，彼が たった今 ノーベル化学賞を受賞したということを知らせる電話を受けたときであった。 理由：Cram はじゅうたん洗いの職にあるので。スウェーデン科学アカデミーは，間違った人を呼び出してしまったのだった。UCLA 化学者の Donald J.Cram はこの混乱を聞いたあと，皮肉って言った：「じゅうたん洗いには何かの化学が関係しているからね」と。

[Ex. 7][2)]　Sustainable Energy System

The world must switch to energy systems that respect the atmosphere's limited capacity for absorbing carbon dioxide while minimizing the exploitation of fossil fuels that cannot be replaced.　If countries incorporate climate change concerns into their development priorities, then the new transport and industrial systems of the 21st century will have a much lighter overall impact on the planet.

　［注］　while 一方で，する間，minimize 最小化する，exploitation 利用，fossil fuel 化石燃料，incorporate A into B：A を B に取り入れる，if ―, then ～：もし―なら，(そのときは)～，impact 影響

　［解説］　頭から順繰りに，アンダーラインごとに対応してみよう。

<u>The world must switch to energy systems</u>　<u>that respect the atmosphere's limited capacity for absorbing carbon dioxide</u>　<u>while minimizing the exploitation of fossil fuels that cannot be replaced.</u>　<u>If countries incorporate climate change concerns into their development</u>

priorities, then the new transport and industrial systems of the 21st century will have a much lighter overall impact on the planet.

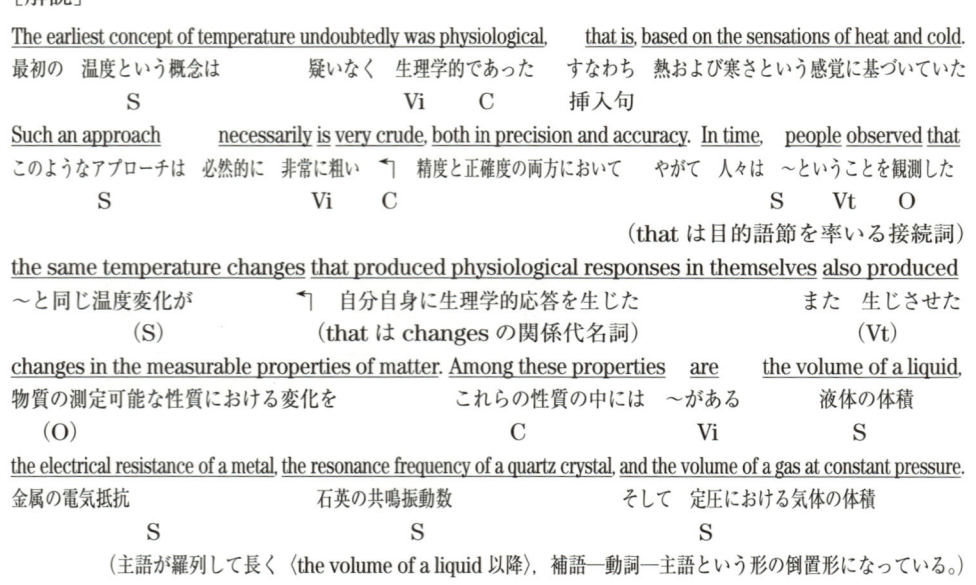

> 和訳例

持続可能なエネルギーシステム

　世界はエネルギーシステムを転換しなければならない。すなわち，大気が CO_2 を吸収する能力には限りのあることを考慮し，一方で，代替できない化石燃料の利用を最小限にするような。もし国々がその開発優先に気候変化の問題を取り入れたなら，21世紀の新たな輸送と産業システムは，この惑星への全般的な影響をずっと軽くするであろう。

[Ex. 8] 6) Temperature

1. The earliest concept of temperature undoubtedly was physiological, that is, based on the sensations of heat and cold. Such an approach necessarily is very crude, both in precision and accuracy. In time, people observed that the same temperature changes that produced physiological responses in themselves also produced changes in the measurable properties of matter. Among these properties are the volume of a liquid, the electrical resistance of a metal, the resonance frequency of a quartz crystal, and the volume of a gas at constant pressure.

　　[注]　concept 概念，undoubtedly 疑いなく，physiological 生理学の，that is（挿入句）すなわち，sensation 感覚，necessarily 必然的に，crude 粗い，生の，precision 精度，accuracy 正確度，in time やがて，間に合って，response 応答，measurable 測定可能な，resonance 共鳴

[解説]

The earliest concept of temperature undoubtedly was physiological,　　that is, based on the sensations of heat and cold.
最初の　温度という概念は　　　疑いなく　生理学的であった　　すなわち 熱および寒さという感覚に基づいていた
　　　　　　S　　　　　　　　　　　　　　　Vi　　　C　　　　挿入句

Such an approach　　necessarily is very crude, both in precision and accuracy.　In time,　people observed that
このようなアプローチは 必然的に 非常に粗い ← 精度と正確度の両方において　やがて　人々は　～ということを観測した
　　　　S　　　　　　　　　　Vi　　C　　　　　　　　　　　　　　　　　　　　　　S　　Vt　　O
　　　　　　　　　　　　　　　　　　　　　　　　　　　　　　　　　　（that は目的語節を率いる接続詞）

the same temperature changes that produced physiological responses in themselves also produced
～と同じ温度変化が　　　　←　自分自身に生理学的応答を生じた　　　　　　　　　　また　生じさせた
　　　（S）　　　　　　（that は changes の関係代名詞）　　　　　　　　　　　　　　（Vt）

changes in the measurable properties of matter. Among these properties　are　the volume of a liquid,
物質の測定可能な性質における変化を　　　　　　　これらの性質の中には　～がある　　液体の体積
　　　　（O）　　　　　　　　　　　　　　　　　　　　C　　　　　　　Vi　　　　　S

the electrical resistance of a metal, the resonance frequency of a quartz crystal, and the volume of a gas at constant pressure.
金属の電気抵抗　　　　　　　　石英の共鳴振動数　　　　　　　そして 定圧における気体の体積
　　　S　　　　　　　　　　　　　　S　　　　　　　　　　　　　　　　S

　　　　　　　（主語が羅列して長く〈the volume of a liquid 以降〉，補語―動詞―主語という形の倒置形になっている。）

和訳例

温 度

　温度という最初の概念は，疑いなく生理学的なもの，すなわち，熱い，冷たいという感覚に基づいていた。このようなアプローチ（取り掛かり，方法）は必然的に精度でも，正確度でも非常に粗いものである。やがて，人々は，自分の生理学的な応答をもたらしたのと同じ温度変化が，また，物質の測定可能な性質の変化を生じるということを観察した。これらの性質の中には，液体の体積，金属の電気抵抗，石英結晶の共鳴振動数，および定圧での気体の体積などがある。

2. Each of these properties can provide the basis for an operational definition of a temperature scale. For example, the Celsius temperature θ is defined by the equation, $\theta = 100 (X_\theta - X_0)/(X_{100} - X_0)$, in which X_θ is the value of the property at temperature θ, X_0 is the value of the property at the temperature of a mixture of ice and water at equilibrium under a pressure of 1 atm (1.0135×10^5 Pa), and X_{100} is the value of the property at the temperature of an equilibrium mixture of water and steam under a pressure of 1 atm

　［注］　operational 操作の，使用可能な，definition 定義，equilibrium 平衡

和訳例

　これらの性質の1つ1つは，温度スケールの使用可能な定義に対してその基礎を提供できる。例えば，セルシウス温度 θ は，式，$\theta = 100 (X_\theta - X_0)/(X_{100} - X_0)$，で定義される。ここで，$X_\theta$ は温度 θ でのその性質の値，X_0 は氷と水の混合物の温度における，1気圧（1.0135×10^5 Pa），平衡でのその性質の値，X_{100} は1気圧下，水と蒸気の平衡混合物の温度における，その性質の値である。

[Ex. 9] [6]　Equilibrium

1. The natural tendency of systems to proceed toward a state of equilibrium is exhibited in many familiar forms. When a hot object is placed in contact with a cold object, they reach a common temperature. We describe the change by saying that heat has flowed from the hot object to the cold object. However, we never observe that the two objects in contact and at the same temperature spontaneously attain a state in which one has a high temperature and the other a low temperature.

　［注］　tendency 傾向，in contact with と接触している，〜 in contact （互いに）接触している〜，object 物体，common 共通の，spontaneously 自然に，自発的に，attain 到達する，one…the other （2者の内）一方…他方．

和訳例

平 衡

　系が平衡状態へ向かう自然の傾向は，多くの見慣れた形で示される。熱い物体が冷たい物体

と接して置かれるとき，それらは共通の温度に到達する．私たちはこの変化を，熱い物体から冷たい物体へ熱が流れたと言って表す．しかし，接触していて，同じ温度にある2つの物体が，一方が高い温度に他方が低い温度をもつような状態に，自発的に到達するようなことを，私たちは決して観測しない．

2. Similarly, if a vessel containing a gas is connected to an evacuated vessel, the gas will effuse into the evacuated space until the pressures in the two vessels are equal. Once this equilibrium has been reached, it never is observed that a pressure difference between the two vessels is produced spontaneously. Solutes diffuse from a more concentrated solution to a more dilute solution; concentration gradients never develop spontaneously. Magnets spontaneously become demagnetized; their magnetism never increases spontaneously. When a concentrated protein solution, such as egg white, is placed in a vessel of boiling water, the egg white coagulates, but we never observe that coagulated egg white at the temperature of boiling water returns spontaneously to a liquid state.

[注] vessel 容器，evacuate 空にする，脱気する，effuse 流出する，solute 溶質，diffuse 拡散する，once ～: 一旦～すると，concentrated 濃縮した，濃い，dilute 希釈した，薄い，gradient 勾配，develop 発現する，demagnetize 消磁する，coagulate 凝集する

[和訳例]
　同様に，気体を含む容器を空の容器とつなぐと，気体は空の空間へ流出して，2つの容器の圧力が等しくなる．一旦平衡になると，2つの容器の間の圧力差が自発的に生じることは，決して観察されない．溶質は，濃い溶液から薄い溶液へ拡散する；濃度勾配が自発的に発現することはない．磁石は自然に消磁するが，磁性が自然に増すことはない．卵白のような濃いタンパク質溶液は沸騰水の容器に入れると，卵白は凝集するが，凝集した卵白が沸騰水温度で自発的に液体状態に戻ることを私たちは決して観察しない．

3. It is desirable to find some common measure (preferably a quantitative measure) of the tendency to change and of direction in which change can occur. In the 1850s, Clasius and Kelvin independently formulated the second law of thermodynamics, and Clausius invented the term entropy S (from the Greek word τροπή, meaning transformation), to provide a measure of the "transformational content," or capacity for change.

[注] quantitative 定量的な，cf. qualitative 定性的な，transformation 変化

[和訳例]
　変化する傾向や変化の起こる方向について何らかの共通の尺度（好ましくは，定量的な尺度）を見つけることが望ましい．1850年代，クラジウスとケルビンは独立に熱力学の第二法則を公式化した．そして，クラウジウスは「変換する容量」すなわち変化のための潜在能力の尺度を与えるために，エントロピーという用語（変化を意味するギリシャ語のτροπήから）を創った．

基礎編 31

[Ex. 10][2)] Hydrogen Bonding

Hydrogen bonding is responsible for the fact that hydroxylic compounds are associated in the liquid state and consequently much less volatile than unassociated liquids of comparable molecular weight. Boiling points of acids are even higher than those of alcohols of comparable molecular weight, as is evident from the comparison of the boiling points of pentanoic acid (MW=102 ; bp=187℃) and n-hexanol (MW=102 ; bp=156℃). The difference is attributable to more effective hydrogen bonding of the former than the latter.

［注］ be responsible for～：～に責任がある，～の原因となる，comparable 比較しうる，同じぐらいの，be attributable to～：～に帰すことができる，～ に基づく，～が原因である，the former, the latter 前者，後者

［解説］ <u>Hydrogen bonding,</u>　<u>is responsible for</u>　　<u>the fact</u>
　　　　水素結合　　は　　　　　～の原因となる　　⤴(次の)事実
　　　　　　S　　　　　　　Vi

<u>that</u>　<u>hydroxylic compounds</u>　<u>are associated</u>　<u>in the liquid state</u>　<u>and consequently</u>
すなわち(conj.)ヒドロキシ化合物　が　　会合している　　液体状態で　　　そして，結果として
(同格節を率いる)　　(S)　　　(Vi)(Vt の受身)

<u>much less volatile</u>　<u>than unassociated liquids</u>　<u>of comparable molecular weight..</u>
はるかに蒸発性が低い　⤴非会合液体よりも　　　⤴　同じような分子量の
C

<u>Boiling points of acids</u>　<u>are even higher</u>　<u>than</u>　<u>those of alcohols of comparable molecular weight,</u>
酸の沸点　は　　　　　さらに高い　　～よりも　⤴　同じような分子量のアルコールのそれ(沸点)
　S　　　　　　　　　Vi　　　　C　　　　　　　(those は boiling points の代名詞)

<u>as</u>　<u>is</u>　<u>evident</u>　<u>from the comparison of the boiling points</u>
沸点の比較から明らかなように
(conj.) Vi　　C

<u>of pentanoic acid(MW=102 ; bp=187℃)and n-hexanol(MW=102 ; bp=156℃).</u>
⤴ペンタン酸(MW=102 ; bp=187℃)と n-ヘキサノール(MW=102 ; bp=156℃)の

<u>The differenc is attributable to</u>　<u>more effective hydrogen bonding</u>　<u>of the former than the latter.</u>
この違い　は　～に帰すことができる　⤴　より効果的な水素結合　　　　⤴　後者よりも前者の
　　S　　　　　Vi　　　　C

[和訳例]

水素結合

水素結合は，次の事実の原因となる。すなわち，ヒドロキシ化合物が液体状態で会合している結果，同じような分子量の非会合液体よりもはるかに蒸発性が低いということの。酸の沸点は，同じような分子量のアルコールのそれ(沸点)よりもさらに高い。ペンタン酸(分子量102, 沸点 187℃)と n-ヘキサノール(分子量102, 沸点 156℃)の沸点の比較から明らかなように。この違いは，後者(n-ヘキサノール)よりも前者(ペンタン酸)の より効果的な水素結合に帰すことができる。

[Ex.11][2] Surface-active Agent

Most of the surface-active agents are a combination of water-attracting or hydrophilic groups on one end of the molecule, and water-repelling or hydrophobic groups on the other.

[注] combination of A and B（combination of A with B）AとBの組み合わせ
or「または」本来の意味よりも，言い換えで「すなわち」の意味で使われることが多い。one（～），the other（～）：(2者の内)一方，他方　cf. one（～），another（～）：（多くの内）1つの，もう1つの（別の）

[解説]

Most of the surface-active agents	are	a combination	of water-attracting or hydrophilic groups
表面活性剤の多くは	である	組合わせ ⌐	水吸引性 すなわち 親水性のグループ(基)の
S	Vi	C	

on one end of the molecule,	and	water-repelling or hydrophobic groups	on the other.
⌐分子の一端に	および	水反発性すなわち疎水性のグループ(基)	⌐他端に
	conj.		(endが省略されている)

[和訳例]

表面活性剤
　　表面(界面)活性剤の多くは，分子の一端に水吸引性すなわち親水性のグループ(基)と，他端に水反発性すなわち疎水性のグループ(基)との組合せである。(内容については，Ex. 38も参照されたい。)

[Ex. 12][2]　Combustible Liquid

Concerning combustible liquids, it must be borne in mind that it is the vapor of the liquid which catches fire, or, when mixed with air, causes an explosion. The flash point is defined as the lowest temperature at which the liquid gives off vapor near the surface in sufficient quantity to form an inflammable mixture with the air. The lower the flash point the more likely the liquid will be set on fire by a flame or a hot surface. If the concentration of the vapor in the air is within certain limits, the flame started by ignition is propagated through the entire volume of the mixture of air and vapor. The more the composition of the mixture approaches the stoichiometric composition, combustion proceeds with such speed and violence that it causes an explosion. The explosive limits are the minimum and maximum concentrations of the vapor in air beyond which propagation of the flame does not occur when the mixture is in contact with a source of ignition. The explosive limits are generally in the zone of relatively low concentration of the vapor. For this reason, an empty flask or drum which has contained an inflammable liquid is more dangerous than if full, the emptied container being more apt to be filled an explosive mixture of vapor and air.

[注]　concerning に関して，combustible 可燃性の，flash point 引火点，borne in mind 心に留める，give off 発する，sufficient to に十分の，inflammable = flammable 可燃性

の (cf. nonflammable 不燃性の), the —(比較級), the 〜(比較級)：—であればあるほど，より〜，set on fire 着火する，concentration 濃度，propagate 伝播する，stoichiometric 化学量論的な，explosive limit 爆発限界，composition 組成，such — that〜：〜ほどの—，dangerous 危険な，apt to しがちである

[解説]

Concerning combustible liquids, it must be borne in mind that it is the vapor of the liquid which catches
可燃性液体に関して，　　　心に留めなければならないことは，　着火したり，あるいは空気と混ざったときに爆発を
fire, or, when mixed with air, causes an explosion. The flash point is defined as the lowest
引き起こすのは　その液体の蒸気であるということである。　引火点は，つぎのように定義される。　すなわち，
temperature at which the liquid gives off vapor near the surface in sufficient quantity to form an
最低温度であり，その温度で，その液体が蒸気をその表面近くに放出して，空気と可燃性混合物を形成するのに十分な量になる。
inflammable mixture with the air. The lower the flash point the more likely the liquid will be set on fire
　　　　　　　　　　　　　　　　　引火点が低ければ低いほど，その液体はよりたやすく，発火する。
by a flame or a hot surface. If the concentration of the vapor in the air is within certain limits,
¶ 炎あるいは熱い表面によって　　　　　　空気中の蒸気の濃度が　ある範囲内にあると，
the flame started by ignition is propagated through the entire volume of the mixture of air and vapor.
発火で始まった炎は　　　　　空気と蒸気の混合物の体積全体に　　伝播する。
The more the composition of the mixture approaches the stoichiometric composition, combustion proceeds
混合物の組成が，化学量論的組成に近づけば，近づくほど，　　　　　　　　　　　燃焼は　進行して
with such speed and violence that it causes an explosion. The explosive limits are the minimum and
爆発を引き起こすほどの速さと激しさになる。　　　　　爆発限界は，空気中の　その蒸気の　最少と最大の
maximum concentrations of the vapor in air beyond which propagation of the flame does not occur
濃度であり，　　　　　　　　　　　　　　これを超えると，炎の伝播は起こらない。
when the mixture is in contact with a source of ignition. The explosive limits are generally in the zone
¶ その混合物が発火源と接触したとき，　　　　　爆発限界は，一般に，比較的低い蒸気濃度の領域にある。
of relatively low concentration of the vapor. For this reason, an empty flask or drum which has
　　　　　　　　　　　　　　　　　　　　　　この理由で，　可燃性液体を入れていた　空のフラスコあるいは
contained an inflammable liquid is more dangerous than if full, the emptied container being more apt
ドラム缶の方が，　　　　　　満タンの場合よりも　いっそう危険である。空の容器の方が　蒸気と空気の爆発混合物で
to be filled an explosive mixture of vapor and air.
満たされがちであるので。

[和訳例]

可燃性液体

　可燃性液体に関して，心に留めておかなければならないことは，着火したり，あるいは空気と混ざったときに爆発を引き起こすのは，その液体の蒸気であるということである。引火点は，液体がその表面近くに，空気との可燃性混合物を作るに十分な量の蒸気を発生するような最低の温度と定義される。引火点が低ければ低いほど，その液体は炎あるいは熱い表面によって，よりたやすく着火する。空気中の蒸気の濃度がある範囲内にあると，発火で始まった炎は，空気と蒸気の混合物の体積全体に伝播する。混合物の組成が化学量論的組成に近づくほど，燃焼は爆発を引き起こすほどのスピードと激しさで進行する。爆発限界とは，空気中のその蒸気の最少と最大の濃度であり，それ(その限界)を超えると，その混合物が発火源と接触し

たときも 炎の伝播は起こらない(そのような濃度範囲のことである)。爆発限界は一般に，蒸気の比較的低い濃度の領域にある。この理由で，可燃性液体を入れていた空のフラスコやドラム缶の方が満タンよりも危険である。空の容器の方が 蒸気と空気の爆発混合物で満たされがちであるので。

[Ex. 13][2] Acid Rain

Most acid rain originates from the combustion of fossil fuels that contain sulfur as an impurity. Sulfur on combustion produces sulfur dioxide. In the atmosphere, sulfur dioxide reacts with oxygen to form sulfur trioxide, which will in turn react with water (rain) to form sulfuric acid, which is one of the strong acids, corrosive and destructive

In all accounts, acid rain is blamed. Some research suggests that increased ozone makes trees more vulnerable to acids. Clean-up programs will be expensive and may yield only qualified results. Yet, countries hardest hit are convinced that something, even if costly and limited, must be done to save invaluable lakes and forests. Eighteen governments have resolved to cut their emissions of sulfur dioxide, nitrogen oxides and other pollutants. Many scientists fear that even if acid-causing emissions were eliminated altogether, it would take decades to restore damaged lakes and forests to their original health and productivity.

[注]　originate 生じる，combustion 燃焼，in turn 次いで，corrosive 腐食性，destructive 有害の，accounts 記事，blame 非難する，vulnerable to (の影響を)受けやすい，invaluable 貴重な(価値の付けられないほどに)，resolve 決定する，eliminate 取り除く，restore 取り戻す，productivity 生産性

[解説]　Most acid rain originates from the combustion of fossil fuels that contain sulfur as an impurity.
多くの酸性雨は，化石燃料の燃焼から 生じる。　　　　　硫黄を不純物として含む
Sulfur on combustion produces sulfur dioxide.　In the atmosphere, sulfur dioxide reacts with oxygen to form sulfur trioxide,
硫黄は，燃焼で二酸化硫黄を生じる。　　　大気中では，二酸化硫黄は酸素と反応して三酸化硫黄を生じ，
which will in turn react with water (rain) to form sulfuric acid : one of the strong acids, corrosive and destructive
それは，次いで，水〈雨〉と反応して硫酸を生じる：　これは，強酸の1つで，腐食性，有害である。
In all accounts, acid rain is blamed.　Some research suggests that increased ozone makes trees more
すべての記事で，酸性雨は非難されている。　ある研究が示唆するには，　オゾンの増加は，樹木をさらに酸の影響を
vulnerable to acids.　Clean-up programs will be expensive and may yield only qualified results.　Yet,
受けやすくする。　　一掃計画は，高価につき，限られた結果しか生じない。　　　　　　それでも，
countries hardest hit are convinced　　　 that　something, even if costly and limited, must be done
もっともひどい被害を受けた国々は，確信している，たとえ高価で限られているとしても，何かがなされねばならないと．
to save invaluable lakes and forests.　Eighteen governments have resolved　to cut their emissions of sulfur dioxide,
貴重な湖や森を救うために。　　　　18の政府が決定した，　　二酸化硫黄，酸化窒素類，その他の汚染物質の
nitrogen oxides and other pollutants.　Many scientists fear　that　even if acid-causing emissions were
排出をカットすることを。　　　多くの科学者は恐れる，～ということを　たとえ酸の原因となる排出がすべて取り除かれた
eliminated altogether, it would take decades　to restore damaged lakes and forests to their original health and productivity.
としても，　　何十年もかかるでしょう。　損傷した湖や森に元の健康と生産性を取り戻させるには，

和訳例

酸性雨

　多くの酸性雨は，硫黄を不純物として含む化石燃料の燃焼から生じる。硫黄は，燃焼で二酸化硫黄を生じる。大気中では，二酸化硫黄は酸素と反応して三酸化硫黄を生じ，それは，次いで，水〈雨〉と反応して硫酸を生じる：これは，強酸の1つで，腐食性，有害である。

　すべての記事で，酸性雨は非難されている。ある研究が示唆するには，オゾンの増加は，樹木をさらに酸の影響を受けやすくする。一掃計画は，高価につき，限られた結果しか生じない。それでも，もっともひどい被害を受けた国々は，確信している：たとえ高価について，（結果が）限られているとしても，貴重な湖や森を救うために，何かがなされねばならないと。18の政府が二酸化硫黄，酸化窒素類，その他の汚染物質の排出をカットすることを決定した。多くの科学者が恐れていることは，たとえ酸の原因となる排出がすべて取り除かれたとしても，損傷した湖や森に元の健康と生産性を取り戻させるには，何十年もかかるであろう，ということである。

[Ex. 14][7]　Iron

1. Iron is believed to be the major component of Earth's core. This metal is also the most important material in our civilization. It does not hold this place because it is the "best" metal; after all, it corrodes much more easily than any other metals. Its overwhelming dominance in our society comes from a variety of factors: (1) Iron is the second most abundant metal in the Earth's crust, and concentrated deposits of iron ore are found in many localities, thus making it easy to mine. (2) The common ore can be easily and cheaply processed thermochemically to obtain the metal. (3) The metal is malleable and ductile, whereas many other metals are relatively brittle. (4) The melting point, 1535℃, is low enough that the liquid phase can be handled without great difficulty. (5) By the addition of small quantities of other elements, alloys that have exactly the required combinations of strength, hardness, or ductility for very specific uses can be formed.

　　[注]　corrode 腐食する，overwhelming 圧倒的な，dominance 優位性，abundant 豊富な，ore 鉱石，locality 地方，産地，mine 採掘する，malleable 展性の，ductile 延性の，― enough that ～：～に十分なほど―，十分に―で，～

和訳例

鉄

　鉄は，地球核の主成分と考えられている。この金属はまた，我々の文明社会においても最も重要な材料である。それは，「ベスト」な金属であるという理由でこの地位を保っているのではない；結局は，それは，他のどんな金属よりも容易に腐食するからである。我々の社会におけるその圧倒的な優位性は，いろんな因子から来る：(1)鉄は地殻中2番目に豊富な金属であり，鉄鉱石の高濃度の堆積物が多くの産地にあって，その採掘を容易にする。(2)共通の鉱石は，熱化学的に容易に安価に処理されて，金属を得ることができる。(3)この金属は展性，延性がある。対して，多くの他の金属は比較的もろい。(4)融点1535℃は十分低くて，液相は，それほど大きな困難なく取り扱える程度である。(5)少量の他の金属を加えることで，非常に

特別な用途に対して，強さ，硬さ，あるいは延性の，丁度必要とされる組み合わせを持った合金を形成することができる。

2. The one debatable factor is iron's chemical reactivity. This is considerably less than that of the alkali and alkaline earth elements but is not as low as that of many transition metals. The relatively easy oxidation of iron is a major disadvantage — consider all the rusting automobiles, bridges, and other iron and steel structures, appliances, tools, and toys. At the same time, it does mean that our discarded metal objects will crumble to rust rather than remain an environmental blight forever.

[注] debatable 異論のある，2行目の2つの that は chemical ractivity の代名詞，rust (v, n) 錆びる，錆び，appliance 器具，discard 捨てる，crumble ぼろぼろになる，崩壊する，blight 暗い影，虫害，as low as と同じぐらい低い

[解説] <u>The one debatable factor is iron's chemical reactivity.</u>　<u>This is considerably less than</u>
　　　　1つの異論のある因子は　　鉄の化学的反応性である。　　　　これは〜よりもかなり低い
<u>that of the alkali and alkaline earth elements</u>　<u>but is not as low as that of many transition metals.</u>
　↲　アルカリおよびアルカリ土類元素のそれ　　しかし，多くの遷移金属のそれほど低いことはない。
<u>The relatively easy oxidation of iron</u>　<u>is a major disadvantage —</u>　<u>consider all</u>
鉄の比較的容易な酸化は　　　　　　　主な欠点である。　　　　考えなさい〈次のこと〉すべてを
<u>the rusting automobiles, bridges,</u>　<u>and other iron and steel structures, appliances, tools, and toys.</u>
錆びる車，　　　　　　　橋，　　そして　その他の鉄およびスチール構造物，器具，道具，そしておもちゃを
<u>At the same time, it does mean</u>　　<u>that our discarded metal objects will crumble to rust</u>
同時に，　　　　　それは意味する　　我々の捨てた金属物体は朽ちて錆になるであろうということを
　　　　　　　　　　　　　　that は目的語節を率いる接続詞（〜ということを）
<u>rather than remain an environmental blight forever.</u>
　↲　永久に環境の暗い影のまま残るよりも，むしろ

[和訳例]
　1つ異論のある因子は，鉄の化学反応性である。これは，アルカリおよびアルカリ土類元素の反応性よりはかなり低いが，多くの遷移金属と同じほど低くはない。鉄が比較的容易に酸化することは主な欠点である—錆びる車，橋，その他の鉄やスチール構造物，器具，道具，おもちゃ，などすべてを考えてごらんなさい。同時にそれが意味することは，我々の捨てた金属物体は，永久に環境の暗い影として残るよりは，むしろ，ぼろぼろになって錆になる，ということである。

[Ex.15][8] Octane Rating of Gasoline

The knock or ping heard when an automobile engine is accelerated too rapidly is a warning that conditions for efficient performance of the engine with the particular gasoline used have been exceeded. The knocking tendency of a given gasoline is expressed as the octane number, or the performance in a standard one-cylinder test engine in comparison with that of mixture of two synthetic standard fuels. Isooctane

(2, 2, 4-trimethylpentane), which detonates only at high compression and was superior to any gasoline known in 1927 when the rating was introduced, was assigned the octane rating of 100, and *n*-heptane, which is particularly prone to knocking, was given the rating 0. The octane number of a fuel is the percent of isooctane in a blend with *n*-heptane that has the same knocking characteristics as the fuel under examination. Investigation of pure synthetic hydrocarbons has shown that in the alkane series octane number decreases as the chain is lengthened and increases with chain branching.

[注]　knock トントンという音，ping ヒューという音，warning 警告，particular 特定の，in comparison with と比較して，detonate 爆発する，compression 圧縮比，prone to しやすい

[解説]　次の Ex. 16, Ex. 17 とともに，古いが，著者らが学生時に使った，著名なフィーザーのテキストから引用した。もちろん内容はそのまま生きている。

和訳例

ガソリンのオクタン価

　自動車エンジンが急加速された時に聞かれるトントンとかヒューという音は，用いた特定のガソリンでのそのエンジンの有効な運転性能の条件が限界を越えたことの警告である。あるガソリンのノッキングの傾向はオクタン数であらわされる。すなわち，ある標準の 1 シリンダー（気筒）テストエンジン中での，2 つの合成標準燃料の混合物と比較した運転性能のことである。イソオクタン (2, 2, 4-トリメチルペンタン) は，高い圧縮比でのみ爆発し，（オクタン）価が導入された 1927 年に知られていたどんなガソリンにも勝っていたので，オクタン価 100 が決められ，*n*-ヘプタンはとくにノッキングしやすいため，オクタン価 0 が与えられた。ある燃料のオクタン価（数）は，その試験中の燃料と同じノッキング特性を持った *n*-ヘプタンとの混合物中のイソオクタンのパーセントである。純粋な合成炭化水素の研究は，アルカン系列では鎖が長いほどオクタン価は減少し，鎖の枝分かれとともに増加することを示している。

[Ex.16][8] **Dimerization of Isobutene to Isooctene**

The dimerization usually induced by phosphoric or sulfuric acid yields the dimer, isooctene (2, 4, 4-trimethylpentene-1) as the main product, which on hydrogenation gives isooctane (2, 2, 4-trimethylpentane).

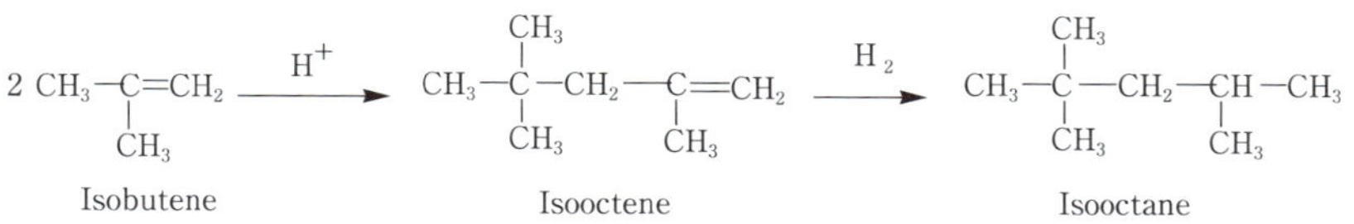

$$2\ CH_3-\underset{CH_3}{\underset{|}{C}}=CH_2 \xrightarrow{H^+} CH_3-\underset{CH_3}{\underset{|}{\overset{CH_3}{\overset{|}{C}}}}-CH_2-\underset{CH_3}{\underset{|}{C}}=CH_2 \xrightarrow{H_2} CH_3-\underset{CH_3}{\underset{|}{\overset{CH_3}{\overset{|}{C}}}}-CH_2-\underset{CH_3}{\underset{|}{CH}}-CH_3$$

　　　Isobutene　　　　　　　　　Isooctene　　　　　　　　　　　Isooctane

Fig. 1. Isooctane from isobutene. （イソブテンからイソオクタン）

The acid-catalyzed dimerization can be interpreted by the following mechanism: (1) a proton attacks the more negative of the two unsaturated carbon atoms to give trimethylcarbonium ion or *tert*-butyl cation ; (2) this ion attacks the more negative center of a second molecule of isobutene ; (3) the resulting dimer ion expels a proton to form isooctene.

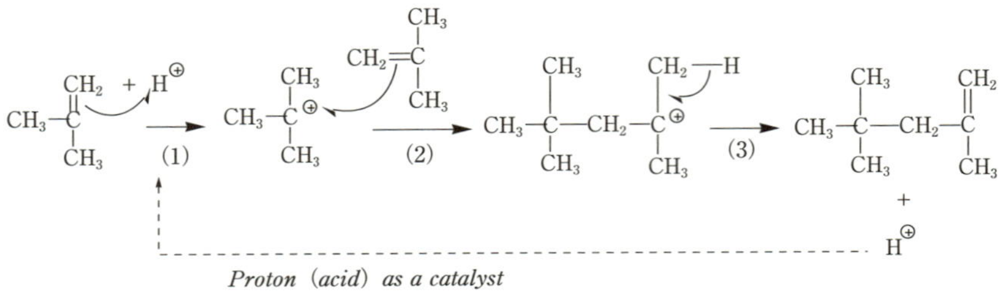

Proton (acid) as a catalyst

Fig. 2. Mechanism of dimerization of isobutene （イソブテンの2量化のメカニズム）

［注］　dimerization 2量化，dimer 2量体（ダイマー），interpret 説明する，expel 放出する

［解説］　ハイオク・ガソリン（上の Ex.15 参照）としてのイソオクタン合成を目的とした，イソブテン2量化の小文である。現在，実際の触媒としては，ゼオライトが用いられる（Ex. 47 参照）。また反応機構（Fig. 2）は，文章とは逆に，電子がカチオンを求核攻撃する，現在ふつうに用いられている形で示した。

[和訳例]

イソブテンのイソオクテンへの2量化

　リン酸あるいは硫酸によって引き起こされる2量化は，通常，主生成物として，ダイマー（2量体）のイソオクテン（2,4,4-トリメチルペンテン-1）を生じる。これは，水素化でイソオクタン（2,2,4-トリメチルペンタン）を与える。

　酸触媒2量化は次の機構で説明することができる：(1)プロトンが2つの不飽和炭素原子のうち陰性の高い方の炭素を攻撃して，トリメチルカルボニウムイオンすなわち *tert*-ブチルカチオンを与える；(2)このイオンは2番目のイソブテン分子の，より陰性の中心（炭素）を攻撃する；(3)生じたダイマーイオンはプロトンを放出して，イソオクテンを生成する。

[Ex.17][8] Tropolones

Some tropolones that occur in nature have been investigated extensively since early times, but the basic structure was not recognized until 1945, when Dewar in England and the Japanese chemist Tetsuo Nozoe in Formosa independently noted that the heptatrienolone structure (1) should have aromatic characteristics, since two equivalent Kekule-like structures are possible, that is, resonance-stabilized (2).

Such natural products as stipitatic acid, a mold metabolite, and hinokitol, a constituent of the essential oil of the Japanese hinoki tree, have aromatic properties but are nonbenzenoid, and the tropolone stuctures suggested by Dewar and Nozoe have been established. The name is derived from that of tropane, the parent substance of the tropane alkaloids, characterized by the presence of a seven-carbon ring. Tropolones exhibit marked phenolic properties: positive ferric chloride test; pK$_a$ values close to 7 (intermediate between phenol and acetic acid); diazo coupling; nitration and bromination. They are resistant to permanganate oxidation, and the ketonic character is masked.

[注]　resonance-stabilized 共鳴安定化した，stipitatic acid スチピタト酸，mold metabolite カビの代謝物，constituent 成分，essential oil 精油

和訳例

トロポロン類

　天然に存在するトロポロン類は，早くから広く研究されてきたが，その基本構造は，1945年まで認識されていなかった。その年，英国のデュワーと，台湾にいた日本人化学者の野副鉄男は，独立に，ヘプタトリエノロン構造(1)は芳香族的性質を持つ筈であることを認めた。というのは，2つの等価なケキュレ状構造が可能，すなわち，(2)のように共鳴安定化されているからである。

　カビの代謝物であるスチピタト酸や，日本のヒノキの精油成分のヒノキトールは芳香族の性質を持つが，非ベンゼン核である。そして，デュワーと野副の示唆したトロポロン構造が確立されたのである。名前は，トロパンアルカロイドの親物質，トロパン，に由来する，これは7炭素環の存在に特徴付けられる。トロポロンは顕著なフェノールの性質を示す：塩化第2鉄($FeCl_3$)テストが陽性；7に近いpKa値（フェノールと酢酸の中間）；ジアゾカップリング；ニトロ化とブロモ化(を受ける)。それらは過マンガン酸塩による酸化を受けず，またケトンの性質もかくされている(示さない)。

[Ex.18][9] Chirality

Chirality (handedness; left or right) is an intrinsic universal feature of various levels of matter. Molecular chirality plays a key role in science and technology. In particular, life depends on molecular chirality, in that many biological functions are inherently dissymmetric. Most physiological phenomena arise from highly precise molecular interactions in which chiral host molecules recognize two enantiomeric guest molecules in different ways. There are numerous examples of enantiomer effects which are frequently dramatic. Enantiomers often smell and taste differently. The structural difference between enantiomers can be serious with respect to the actions of synthetic drugs. Chiral receptor sites in the human body interact only with drug molecules having the proper absolute configuration, resulting in marked differences in the pharmacological activities of enantiomers.

［注］　chirality(handedness)キラリティ(掌性)，　enantiomer 鏡像体
［解説］　野依良治氏のノーベル賞講演(2001)"Asymmetric Catalysis: Science and Opportunities(不斉触媒：科学と機会)"の冒頭からである。

Chirality(handedness; left or right) is an intrinsic universal feature of various levels of matter. Molecular chirality plays
キラリティ(掌性，左または右)は，さまざまなレベルの物質に固有の普遍的特徴である。　分子キラリティは科学と技術
a key role in science and technology. In particular, life depends on molecular chirality, in that many biological functions are
において鍵となる役割を演じる。　　とくに，　生命は 分子キラリティに依存する。　多くの生物学的機能は 本来 反対称である
inherently dissymmetric. Most physiological phenomena arise from highly precise molecular interactions　in which chiral host
という点で，　　多くの生理学的現象は　非常に精密な分子相互作用から生じる。　　　　　そこでは，キラルなホスト
molecules recognize two enantiomeric guest molecules in different ways. There are numerous examples of enantiomer effects
分子が　2つの鏡像ゲスト分子を　違った方法で認識する。　　　　　鏡像体効果には多くの例があり，
which are frequently dramatic. Enantiomers often smell and taste differently. The structural difference between enantiomers can
それらはしばしば劇的である。　鏡像体はしばしば違った匂いと味がある。　鏡像体の構造の違いは，
be serious with respect to the actions of synthetic drugs. Chiral receptor sites in the human body interact only with drug molecules
合成医薬の作用に関して　深刻になることもある。　　　　人体のキラルな受容体部位は，　　正しい絶対立体構造をもった
having the proper absolute configuration, resulting in marked differences in the pharmacological activities of enantiomers.
医薬分子とのみ相互作用する，　　　　その結果，鏡像体の薬理作用に著しい違いを生じる。

［和訳例］

キラリティ

　　キラリティ(掌性，左または右)は，さまざまなレベルの物質に固有の普遍的特徴である。分子キラリティは科学と技術において鍵となる役割を演じる。とくに，多くの生物学的機能は本来反対称であるという点で，生命は分子キラリティに依存する。多くの生理学的現象は 非常に精密な分子相互作用から生じる。そこでは，キラルなホスト分子が2つの鏡像ゲスト分子を違った方法で認識する。鏡像体効果には多くの例があり，それらはしばしば劇的である。　鏡像体はしばしば違った匂いと味がある。　鏡像体の構造の違いは，合成医薬の作用に関して 深刻になることもある。人体のキラルな受容体部位は，正しい絶対立体構造をもった医薬分子とのみ相互作用する，その結果，鏡像体の薬理作用に著しい違いを生じる。

[Ex. 19][2] Virus

The viruses are one kind of giant molecules which have very interesting properties. They have the power of self-duplication—that is, the power of causing other molecules identical with themselves to be formed when they are in the right environment. A disease such as measles results from the formation of a great many measles-virus molecules in the human body which has been infected by a few of these molecules. Another property which virus molecules have, in common with ordinary small molecules, is the ability to form crystals.

［注］　self-duplication 自己複製，cause ～ (to)—：～ に(を)—させる，measles はしか，infect 感染させる

［解説］

The viruses are one kind of giant molecules　　which have very interesting properties.
ウイルスは，一種の巨大分子であり，　　　　　非常に面白い性質を持っている。
They have the power of self-duplication — that is, the power of　causing other molecules identical with themselves to be formed
それらは，自己複製の能力を持つ　　すなわち，能力である　　↰自分と同じ 別の分子を生成させる
when they are in the right environment.　A disease such as measles results from
↰ それらが正しい環境にあるときに，　　　はしかのような病気は，　～から生じる
the formation of a great many measles-virus molecules in the human body
↰ 人体に 非常に多くのはしかウイルス分子が生成すること
which has been infected by a few of these molecules.
↰ ほんの少しのこれらの(はしかのウイルスの)分子に感染した
Another property which virus molecules have,　　in common with ordinary small molecules,
ウイルス分子が持っているもう一つの性質は，　　↰ 普通の小分子と共通して，
is the ability to form crystals.
結晶を形成する能力である

［和訳例］

ウイルス

　ウイルスは，一種の巨大分子であり，非常に面白い性質を持っている。それらは，自己複製の能力を持つ—すなわち，それらが正しい環境にあるときに，自分と同じ 別の分子を生成させる能力である。はしかのような病気は，ほんの少しのはしかのウイルス分子に感染した人体に非常に多くのはしかウイルス分子が生成することから 生じる。ウイルス分子が，普通の小分子と共通して，持っているもう１つの性質は，結晶を形成する能力である。

[Ex. 20] [10)]Carbohydrates

1. Carbohydrates are found in every living organism. The sugar and starch in food and the cellulose in wood, paper, and cotton are nearly pure carbohydrate. Modified carbohydrates form part of the coating around living cells; other carbohydrates are found in the DNA that carries genetic information, and still others are used as medicines.

［注］　modified 修飾された，genetic 遺伝の

［和訳例］

炭水化物

　炭水化物は，あらゆる生物体中にある。食物中の糖とでんぷん，木，紙，綿中のセルロース

は，ほとんど純粋な炭水化物である。修飾炭水化物は生細胞を覆っているものの一部を形作り，他の炭水化物は遺伝情報を運ぶ DNA 中にあり，さらに他のものは医薬に用いられる。

2. The word carbohydrate derives historically from the fact that glucose, the first carbohydrate to be obtained pure, has the molecular formula $C_6H_{12}O_6$ and was originally thought to be a "hydrate of carbon", $C_6(H_2O)_6$. This view was soon abandoned, but the name persisted. Today, the term carbohydrate is used to refer loosely to the broad class of polyhydroxylated aldehydes and ketones commonly called sugars.

   [注]　abandon 捨てる，persist 固執する，持続する，refer to を指す，言及する，polyhydroxylated ポリヒドロキシル化された

   和訳例

   　　炭水化物という言葉は，歴史的には，純粋に得られた最初の炭水化物であるグルコースが $C_6H_{12}O_6$ という分子式をもっていて，初めは「炭素の水和物」$C_6(H_2O)_6$ であると考えられた事実に由来する。この見方はまもなく捨てられたが，名前は持続している。今日，炭水化物という用語は，ゆるく，ふつう糖と呼ばれている広いクラスのポリヒドロキシアルデヒドおよびケトンを指すのに，使われている。

3. Carbohydrates are made by green plants during photosynthesis, a complex process in which sunlight provides the energy to convert CO_2 and H_2O into glucose. Many molecules of glucose are then chemically linked for storage by the plant in the form of either cellulose or starch. It has been estimated that more than 50% of the dry weight of the earth's biomass — all plants and animals — consists of glucose polymers. When eaten and then metabolized, carbohydrates provide the major source of energy required by organisms. Thus, carbohydrates act as the chemical intermediaries by which solar energy is stored and used to support life.

$$6\ CO_2 + 6\ H_2O \xrightarrow{\text{Sunlight}} 6\ O_2 + C_6H_{12}O_6 \longrightarrow \text{Cellulose, Starch}$$

   [注]　photosynthesis 光合成，complex 複雑な，biomass バイオマス（生物資源），intermediary 仲介者（物）

   和訳例

   　　炭水化物は，緑色植物によって光合成で作られる，すなわち，太陽光がエネルギーを供給して，CO_2 と H_2O をグルコースに変換する複雑な過程で作られる。その後，多くのグルコース分子が化学的に結合されて，セルロースあるいはでんぷんの形で植物によって貯蔵される。地球の生物資源—すべての動植物—の乾燥重量の 50% 以上がグルコースの重合体から成ると推定されている。食べて，次いで消化された炭水化物は，有機体に必要な主なエネルギー源を提供する。このように，炭水化物は，太陽エネルギーが蓄えられ，そして生命維持に用いられるための化学的仲介物として働く。

4. Carbohydrates are generally classed into two groups: simple and complex. Simple sugars, or monosaccharides, are carbohydrates like glucose and fructose that can't be converted into smaller sugars by hydrolysis. Complex carbohydrates are composed of two or more simple sugars linked together. Sucrose (table sugar), for example, is a disaccharide made up of one glucose molecule linked to one fructose molecule. Similarly, cellulose is a polysaccharide made up of several thousand glucose molecules linked together. Hydrolysis of polysaccharides breaks them down into their constituent monosaccharide units.

［注］ convert into に変換する，constituent 成分の，構成する，consist of 〜，made up of 〜，composed of 〜：〜から成る，monosaccharide モノサッカライド〈単糖〉，disaccharide ジサッカライド〈2糖〉，polysaccharide ポリサッカライド（多糖）

和訳例

　　炭水化物は一般に2つのグループに分類される：単純な糖と複雑な糖である。単純な糖すなわちモノサッカライドは，グルコースやフルクトースのような炭水化物で，加水分解によってもっと小さい糖に変換されることができない。複雑な炭水化物は，2つ以上の単糖が結合したものから成る。例えば，スクローズ（食卓砂糖）は，1つのグルコース分子が1つのフルクトース分子に結合したものから成るジサッカライド（2糖）である。同様に，セルロースは，数千のグルコース分子が互いに結合したものから成るポリサッカライド（多糖）である。ポリサッカライドの加水分解は，それらの成分モノサッカライド単位へと分解する。

5. Monosaccharides are further classified as either aldoses or ketoses. The -ose suffix is used as the family name ending for carbohydrates, and the aldo- and keto- prefixes identify the nature of the carbonyl group, whether aldehyde or ketone. The number of carbon atoms in the monosaccharide is indicated by using tri-, tetr-, pent-, hex-, and so forth, in the name. For example, glucose is an aldohexose, a six-carbon aldehyde sugar; fructose is a ketohexose, a six-carbon keto sugar: and ribose is an aldopentose, a five-carbon aldehyde sugar. Most of the commonly occurring simple sugars are either aldopentoses or aldohexoses.

［注］ whether A or B, either A or B：AかBかどちらか

和訳例

　　モノサッカライドは，さらに，アルドースかケトースのどちらかに分類される。オースとい

う語尾は，炭水化物に対する族名の末尾として用いられ，アルドやケトの接頭語はカルボニル基を，アルデヒドかケトンか同定するものである。モノサッカライドの炭素原子数は，名前の中に，トリ，テトラ，ペンタ，ヘキサなどによって示される。例えば，グルコースはアルドヘキソースであり，6炭素のアルデヒド糖である。フルクトースはケトヘキソースで，6炭素ケトン糖である。そして，リボースはアルドペントース，すなわち5炭素アルデヒド糖である。通常存在する単糖のほとんどは，アルドペントースかアルドヘキソースのどちらかである。(Ex. 60 も参照されたい。)

[Ex. 21][11] DNA-The Architect of Life

米国化学会(ACS = American Chemical Society)の著した啓蒙書(1981)から，DNAの話「DNA-生命の建築家」を引用しよう。生き生きとした英文が見られる。

1. Inside a human cell nucleus are 46 fine "threads" called chromosomes. Remarkably, each of the 50 trillion cells has a set of these 46 chromosomes. They contain a complete plan for your entire body; much like volumes of instructions in an enormous library devoted to the topics of making and maintaining you. If the instructions coded by the human chromosomes were to be written out — and only four letters are required to write out this code — they would correspond to 1,000 volumes of 1,000 pages each, and each page would contain 1,000 letters.

　Coiled inside the chromosomes are very remarkable, long, thread-like molecules called deoxyribonucleic acid, or simply, DNA. DNA is found in every living thing, except a few primitive viruses.

[注]　trillion [米] 兆(10^{12})，chromosome 染色体，instruction 指示，code 暗号にする

[和訳例]

DNA-生命の建築家

　人の細胞核の中には，染色体と呼ばれる46の細い「糸」がある。注目すべきことに，50兆の細胞それぞれがこれら46の染色体を持っている。それらは，人の身体全体の完全な設計図を含んでいる；ちょうど，人を作り，維持するという話題に捧げられた，巨大な図書館にある何巻もの指示書のように。もし，人の染色体に暗号付けされた指示が書き出されたとすると，—しかも，この暗号を書き出すのにたった4文字だけが必要とされる—それらは，各1,000頁の1,000巻，しかも各1頁が1,000文字を含むものに相当する。

　染色体の中にコイルを巻いているのは，デオキシリボ核酸あるいは単にDNAと呼ばれる，非常に注目すべき，長い糸状の分子である。DNAは，少しの原始的なウイルスを除けば，あらゆる生物に存在する。

2. DNA is the essence of life. With the help of other cellular mechanisms, it can make a copy of itself, and be passed from cell to cell, individual to individual, in literally a biological message from generation to generation.

　Specific segments of the DNA molecule are called genes, and as Mendel theorized more than 100 years ago, it is the genes that carry very precise instructions for different traits such as eye color, insulin production, or hair texture. Groups of genes coded for one specific trait can be compared in a sense to a single page from the 1,000 volumes in DNA's library. An individual gene is like one sentence on the page.

[注] literally 文字通り，be compared to と比較される，と同じぐらいである
it — that ~：~は—である（—を強調）

> 和訳例

　　DNAは生命の本質である。他の細胞の仕組の助けを得て，それは，自身のコピーを作り，細胞から細胞へ，個人から個人へ，文字通り，世代から世代への生物学的メッセージの形で，受け渡される。
　　DNA分子の特定の部分は遺伝子と呼ばれ，メンデルが100年以上も前に理論付けしたように，目の色，インシュリンの生産，あるいは毛の組織のようなさまざまな特徴に対する非常に精密な指示を担っているのが，遺伝子である。ある特定の特徴を暗号化した一群の遺伝子は，ある意味で，DNA図書館の1,000巻からの1頁と同じぐらいである。個々の遺伝子は，その頁の1文のようである。

3. DNA is the part of us that spans generations. It is constantly being replicated — reproduced. And it struggles constantly to maintain the accuracy of the message it will transmit to the next generation of cells, and to the next generation of the species. Events which affect the DNA are crucial to life; mistakes in its duplication play critical roles in shaping the future. To come to even a simple understanding of life's fundamental mechanisms has consumed many decades of work by some of the greatest scientific minds. We are still far from comprehending how a set of instructions encoded by the genes of DNA direct the development of the 50 trillion cells which act together as a human being. Biochemists and other scientists press the search, for only through this continuing effort can we hope to fully understand life.

[注] span generations 何代にも亘る，replicate 複製する，struggle 戦う，crucial 重大な，致命的な，critical 決定的な，be still far from comprehending ~ を理解するにはなおほど遠い

> 和訳例

　　DNAは何代にもわたる我々の一部である。それは常に複製され，再生されている。そして，メッセージの正確さを維持するために絶えず戦い，そのメッセージを次世代の細胞へ，そして次世代の種へと伝達する。DNAに影響を与える出来事は生命に重大である；複製の誤りは未来を形作るのに決定的な役割を演じる。生命の基本的なメカニズムの単純な理解に達することさえも，何人かの偉大な科学者が数十年の研究を費やしている。我々は，DNA遺伝子に暗号化された一組の指示がどのように，人間として一緒に働く50兆の細胞の発生を導くかを，理解するにはなおほど遠い。生化学者および他の科学者は研究を推し進める。この継続的な努力を通してのみ，我々は生命を完全に理解する希望を持てるからである。

4. If we could increase the magnification sufficiently, we could see DNA's individual threads arranged in a long double helix, much like a twisted ladder. And

we would find that the building blocks of DNA are clusters of atoms such as carbon, nitrogen, oxygen, hydrogen, and phosphorus. The strands of DNA are made of two chains of sugar and phosphate. Attached to sugars in each strand are subunits called purine and pyrimidine bases, which correspond in our analogy to the rungs of a ladder.

A series of rungs together with two strands comprise a specific gene for a specific trait. A typical gene has hundreds of pairs of bases. Chemists have named the four common bases Adenine, Cytosine, Guanine, and Thymine. They are arranged in pairs. "A" always goes with "T" and "C" invariably with "G". Therefore, these bases, or rungs, are the alphabet which spell out DNA's messages for maintaining old cells and building new cells.

[注] magnification 倍率, sufficiently 十分に, individual 個々の, twist ねじる, ladder はしご, building block ビルディングブロック（構成単位）, cluster かたまり, strand より糸, sugar 糖, phosphate リン酸エステル, リン酸塩（DNA 構成単位として, このエステルと塩の両者を含む：〔7項の〔訳注〕Fig.3 を参照〕, attach 結合する, rung（はしごの）段

[解説] 一部のみ SVOC 関係を示す。

If we could increase the magnification sufficiently, we could see DNA's individual threads arranged
   (S)   (Vt)   (O)   S   Vt   O   C
in a long double helix, much like a twisted ladder. And we would find   that   the building blocks of DNA
                                                            S   Vt   O(接続詞：～ということ) (S)
are   clusters of atoms such as carbon, nitrogen, oxygen, hydrogen, and phosphorus. The strands of
(Vi)   (C)   S
DNA are made   of two chains of sugar and phosphate. Attached to sugars in each strand are subunits
    Vt の受動態   C   C(Vt の受動態)   Vi   S
called purine and pyrimidine bases, which   correspond in our analogy to the rungs of a ladder.
                                       (S)   (Vi)

[和訳例]

　もし私たちが倍率を十分上げることができたとしたら、私たちは、長い二重らせんに配列していて、ちょうどねじれたはしごのような、DNA の個々の糸を見ることができるであろう。そして、DNA の構成ブロックは炭素、窒素、酸素、水素、そしてリンなどの原子のかたまりであることを見出すだろう。DNA のより糸は、糖とリン酸エステルの2つの鎖からできている。各々の糸の糖に結合しているのは、プリンとピリミジン塩基であり、これらは、私たちの例えでは、はしごの段に相当する。

　2本のより糸とともに、一連のはしごの段は、ある特定の性質に対する特定の遺伝子を構成する。典型的な遺伝子は数百個の塩基対を持っている。化学者は4つの共通の塩基をアデニン(A)、チトシン(C)、グアニン(G)、そしてチミン(T)と名づけた。これらは対になって配列している。「A」はいつも「T」と、「G」はいつも「C」と一緒に（対に）なるのである。したがって、これらの塩基、すなわちはしごの段が、DNA メッセージを書き綴り、古い細胞を維持し、新しい細胞を作るのである。

5. Life's master molecule is responsible for three processes. We will examine each in some detail, as science now understands these processes. The first is

duplication, in which a copy of DNA is made. This must occur whenever a new cell is to be created. In duplicating, DNA reproduces all the volumes in the library of genetic information for use by the new cell.   The other two processes are called transcription and translation, which allow genes to be expressed and build the cell and body components. These processes are performed continuously, and are absolutely necessary for the maintenance of old and new cells.

DNA duplication is exceedingly complex due to both its helical structure and the need for safeguards to prevent errors. Duplication of DNA begins when a special protein first unwinds the two DNA strands and then separates them at the paired bases. Each strand can now serve as a template, or mold, for a new strand. With the bases of each half of the original molecule now exposed, new bases from the surrounding jelly-like substance are free to bump into and join with them. This is not, however, a random process. Only the proper matching base will do, always "A" mating to "T", "C" mating to "G", and so on. This difficult task of selection is the responsibility of an enzyme complex, including DNA polymerase. It checks for the proper bases, removing the wrong ones and assisting the chemical reactions which join the correct ones.

［注］　duplication 複製，transcription 転写，translation 翻訳，allow to させる，exceedingly 非常に，過剰に，complex 複雑な，unwind ほどく，template テンプレート（型板），mold 鋳型，bump into ぶつかる，complex 複合体，錯体，polymerase ポリメラーゼ（重合酵素）

[和訳例]

　　生命の支配分子は３つの過程の任務を果たしている。今の科学がこれらの過程を理解しているままに，それぞれを少し詳細に調べよう。最初は，複製であり，ここではDNAのコピーが作られる。これは，新しい細胞が作られるときはいつも起こらなければならない。複製する際に，DNAは，新しい細胞に用いられるように遺伝子情報という図書館の中にある全集を再生する。他の２つの過程は転写および翻訳と呼ばれ，これらは遺伝子を表現させて，細胞と身体成分を作らせる。これらの過程は連続的に実行され，古い細胞と新しい細胞の維持に絶対的に必要である。

　　DNAの複製は，そのヘリックス（らせん）構造と，エラーを防ぐ保護の必要性のために，非常に複雑である。DNAの複製は，特別なタンパク質がまず２本のDNAより糸をほどいて，対になった塩基のところで，それらを引き離すときに始まる。それぞれの糸は，今度は，新しい糸のためのテンプレートあるいは鋳型として役立つ。元の分子のそれぞれの半分の塩基が，今や，さらされると，まわりのジェリー状の物質から新しい塩基がそれらにぶつかり，仲間入りする。しかし，これはランダム（無差別な）過程ではない。うまく適合する塩基だけができるのである：常に，「A」は「T」に適合し，「C」は「G」に適合する，などである。この困難な選択の仕事は，DNAポリメラーゼを含む酵素複合体の責務である。それは，正しい塩基をチェックし，誤ったものを除き，正しい塩基を仲間に入れる化学反応を助ける。

6. In spite of all of this complexity, duplication of DNA occurs with amazing speed. A well-fed cell of bacteria, E-coli, can duplicate itself in about 20 to 40

minutes. Calculations based on the length of the bacteria indicate that to accomplish this, the DNA must be unwound at the rate of about 5,000 revolutions per minute, twice as fast as the propeller of a small plane spins at cruising speed. New bases must be added at the astonishing rate of nearly 800 per second.

DNA must maintain the cell by processes called transcription and translation.

Transcription is the process of reading and copying the instructions of the DNA into a working plan. Transcription is similar to duplication in that proteins again bind to and unwind the helix; however, this time, only one chain is read by an enzyme, RNA polymerase. The chain's four letter code, A, C, G, and T, is transferred into a new four-letter code, A, C, G, and U, for Uracil, to make a messenger molecule called ribonucleic acid, or RNA.

As the messenger RNA is produced, cellular units, called ribosomes, bind to it, beginning the process of translation into protein. The ribosome can be likened to translators of a foreign language. They have the capability of translating the coded four-letter message, now residing in RNA, into 20 amino-acid alphabet for making proteins, the building blocks of living organisms. Through this mechanical way, the instructions of a gene are expressed. The gene has actively directed the assembly of these amino acids, in the proper sequence, into a protein chain.

> [注]　amazing 驚くべき，well-fed 栄養の十分な，*E-coli* 大腸菌，accomplish 成し遂げる，revolutions per mimute 毎分の回転，astonishing 驚くべき，messenger 伝令，ribosome リボソーム，assembly 集合，sequence 順序

[和訳例]

　　この複雑さすべてにも関わらず，DNA の複製は驚くべき速さで起こる。栄養の十分なバクテリア，大腸菌，はおよそ 20 ないし 40 分で自分を複製できる。バクテリアの長さに基づいて計算すると，これを成し遂げるには，DNA は毎分約 5,000 回転の速度でほどけなければならない。これは小さな飛行機のプロペラが巡航スピードで回転する 2 倍の速さである。新しい塩基は，1 秒当たり 800 個近くの驚くべき早さで加えられているに違いない。

　　DNA は，転写と翻訳と呼ばれる過程で細胞を維持しなければならない。

　　転写は，DNA の指示を読み，作業計画にコピーする過程である。転写は，タンパク質が再びヘリックスに結合して，ほどくという点で，複製に似ている；しかし，今回は 1 本の鎖だけが酵素，RNA ポリメラーゼ，によって読まれる。鎖の 4 つの文字暗号，A，C，G，T，は，新しい 4 つの文字暗号，A，C，G，U（ウラシル），に換えられて，リボ核酸すなわち RNA と呼ばれる伝令分子を作る。

　　伝令 RNA が作られると，リボソームと呼ばれる細胞の単位がそれに結合し，タンパク質への翻訳の過程を始める。リボソームは外国語の翻訳機にたとえることができる。それらは，今は RNA に収まっている暗号の 4 文字メッセージを，生物体のビルディング・ブロックであるタンパク質のための 20 個のアミノ酸のアルファベットに翻訳する能力を持っている。この機械的な方法により，遺伝子の指示が表現される。遺伝子は，積極的に，これらのアミノ酸を，正しい順序で，タンパク質の鎖への集合を指示しているのである。

7. These same life processes — duplication, transcription, and translation of DNA — are shared by creatures throughout the world. And, because of DNA, the elephant

and the bacterium, the towering redwood, each blade of grass, and mankind, all are related: differing only because of variations in the arrangement of the four chemical bases along the DNA spiral.

［注］ redwood セコイア（アメリカスギ）．

和訳例

　これらの同じ生命過程—DNA の複製，転写，および翻訳—は世界中の生命体で共有されている。そして，DNA の故に，象も，バクテリアも，そびえ立つセコイア（アメリカスギ）も，草のひとつひとつの葉も，人も，すべてが関連している：違うのは，ただ，DNA らせんに沿った 4 つの化学塩基（A, T, G, C）の配列の変化の故である。

RNA では，DNA の *deoxyribose* と T が，それぞれ *ribose* と U に換わる：

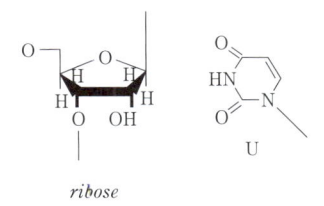

［訳注］　Fig. 3. Partial structure of DNA and RNA
　　　　　（DNA, RNA の部分構造式：破線は水素結合を示す）

[Ex. 22]¹⁰⁾　Natural Rubber

1. Rubber is originally a naturally occurring alkene polymer produced by more than 400 different plants. The major source, however, is the so-called rubber tree, Hevea Brasiliensis, from which the crude material is harvested as it drips from a slice made through the bark. The name rubber was coined by Joseph Priestley, the discoverer of oxygen and early researcher of rubber chemistry, for the simple reason that one rubber's early uses was to rub out pencil marks on paper.

　注：crude 生の，粗の，harvest 収穫する，slice 切れ目，bark 木皮，coin 造る，rub 擦る

和訳例

天然ゴム

ゴムは，もともとは，400 を超えるさまざまな植物によって作られた，天然由来のアルケンポリマーである。しかし，その主な源はいわゆるゴムの木，ヒビア・ブラジリエンシスであり，これから生原料が，皮につけた切れ目から滴り出るときに収穫される。ゴムという名前は，酸素の発見者で，ゴム化学の初期研究者であった，ヨセフ・プリーストリーによって造られた，その単純な理由は，ゴムの初期の利用の1つが紙の上の鉛筆の跡を擦り (rub) とること (消しゴム) であった。

2. Unlike polyethylene and other simple alkene polymers, natural rubber is a polymer of conjugated diene, isoprene, or 2-methylbuta-1, 3-diene. The polymerization takes place by 1, 4-addition of isoprene monomers to the growing chain, leading to a polymer that still contains double bonds spaced regularly four-carbon intervals. The double bonds have *cis*- or *Z*-stereochemistry.

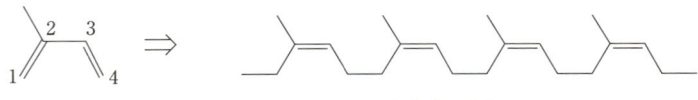

cis-1, 4-polyisoprene

[注]　conjugated 共役した，growing chain 成長している鎖，Z-(ドイツ語 zusammen- 一緒に)　*cis*- と同じ意味

和訳例

ポリエチレンや他の単純アルケンのポリマーと違って，天然ゴムは共役ジエン，イソプレン，すなわち 2-メチルブタ-1, 3-ジエンのポリマーである。重合は，成長鎖へのイソプレンモノマーの 1, 4-付加によって起こり，なお二重結合を規則的に 4 炭素間隔毎に含むポリマーを導く。二重結合はシス-または Z-立体化学をもつ。

3. Crude rubber, called latex, is collected from the tree as an aqueous dispersion that is washed, dried, and coagulated by warming in air to give a polymer with chains that average about 5,000 monomer units in length and have molecular weights of 200,000 to 500,000 amu. The crude polymer is too soft and tacky to be useful until it is hardened by heating with elemental sulfur, a process called vulcanization.

[注]　dispersion 分散(液)，coagulate 凝集する，average (*v*) 平均して〜である，amu = atomic mass unit 原子質量単位，tacky べとべとした，粘着性の，vulcanization 加硫

和訳例

生ゴムは，ラテックスと呼ばれ，木から水分散液として集められ，これは洗われ，乾燥されて，空気中暖めることで凝集され，平均約 5,000 のモノマー単位の長さで，分子量 200,000 ないし 500,000amu の鎖のポリマーが得られる。この生 (未加工) のポリマーは，柔らか過ぎ，べとべとし過ぎていて，元素硫黄と加熱して，加硫といわれる過程で硬められるまでは，役に立

たない。

4. By mechanisms that are still not fully understood, vulcanization cross-links the rubber chains together by forming carbon—sulfur bonds between them, thereby hardening and stiffening the polymer. The exact degree of hardening can be varied, yielding material soft enough for automobile tires or hard enough for bowling balls (ebonite).

［注］ cross-link 架橋する，yield 生じる，産する

和訳例

　　まだ完全には分かっていないメカニズムで，加硫は，鎖間に炭素―硫黄結合を作ることによって，ゴム鎖を互いに架橋する，そしてポリマーを硬く，強くする。硬化の正確な程度は変えることができて，自動車タイヤ用の十分柔らかい材料あるいはボーリング・ボール用の十分硬い材料(エボナイト)を産する。

5. The remarkable ability of rubber to stretch and then contract to its original shape is due to the irregular shapes of the polymer chains caused by the double bonds. These double bonds introduce bends and kinks into the polymer chains, thereby preventing neighboring chains from nestling together into tightly packed, semicrystalline regions. When stretched the randomly coiled chains straighten out and orient along the direction of the pull but are kept from sliding over each other by the cross-links. When the stretch is released, the polymer reverts to its original random state.

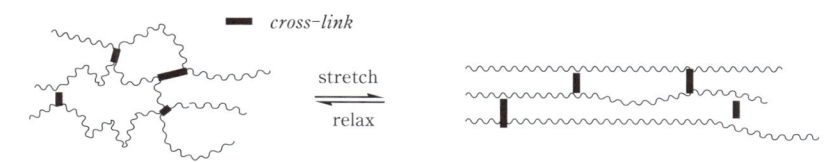

［注］ bend 曲がり，kink ねじれ，nestle(巣を作るように)擦り寄る，release 開放する，revert 戻す，prevent — from ~, keep — from ~ : —を~から防ぐ，妨げる

和訳例

　　ゴムが伸びて，元の形に縮むという素晴らしい能力は，二重結合によって引き起こされるポリマー鎖の不規則な形による。これらの二重結合は，ポリマー鎖中に曲がりやねじれを導入し，それによって，近接鎖が互いに擦り寄って，きつく詰まった，半結晶性領域になるのを妨げる。伸びたとき，ランダムにコイルした鎖はまっすぐ伸びて，引っ張り方向に配向するが，架橋によってお互いが滑りあうのが妨げられる。伸びがゆるめられると，ポリマーは元のランダム状態に戻る。(Ex. 63 参照)

[Ex. 23][12] Nylon

少し長くなるが，まとまった内容のテキストにも，挑戦しよう。

1. Nylons are synthesized either by combination (polycondensation) of diamines and dicarboxylic acids or by ring-opening polymerization of lactams: e, g.,

The amide functional group, which acts as the link holding the units of the nylon chain together, is also a functional group of great importance to biochemistry. The linking group of proteins is also the amide function and so in this way nylon is a synthetic polymer that is a biological analog. Indeed, the natural polymer that nylon was competing against, used earlier for women's stockings, silk, also contains the amide functional group to hold the units of the chain together. Nylon was first called artificial silk.

[注] either A or B: A または B どちらか（どちらでも），combination 結合，polycondensation 重縮合（小分子，ここでは H_2O 分子，が取れる縮合反応の繰り返し），ring-opening polymerization 開環重合，functional group 官能基，biological analog 生物類似体，compete against と対抗する

和訳例

ナイロン

ナイロンは，ジアミンとジカルボン酸の結合（重縮合）か，ラクタムの開環重合のどちらかで合成される。

|図：ヘキサメチレンジアミンとアジピン酸の重縮合，および ε-カプロラクタムの開環重合による，それぞれ，ナイロン6,6，ナイロン6の合成|

アミド官能基は，ナイロン鎖の単位を繋ぎ合わせる連結器具（結合）として働く，とともに生化学に大きな重要性を持つ官能基でもある。タンパク質の結合基もアミド官能基であり，したがってこのような点でナイロンは生物類似の合成高分子といえる。実際，ナイロンと張りあい，以前から女性のストッキングとして使われていた天然高分子，絹，もまたアミド官能基を含み，鎖の単位を結合している。ナイロンは最初，人造の絹（人絹）と呼ばれた。

2. There is hydrogen on the nitrogen of each amide group linking together the carboxylic acids and amines in nylon. It also links the amino acid-derived units in

silk. This hydrogen bound to nitrogen plays an essential role in the overall structures of all nylons and in proteins of all varieties. In the proteins in silk, this hydrogen is most notably involved in the formation of β-sheets, which hold the protein chains together in an arrangement that is responsible for the strength and fibrous properties of the silk. Similarly in nylons this hydrogen—which is responsible for very strong interactions among the polymer chains—is also responsible for the fiber-forming properties of the nylons. In both silk and nylon the nitrogen-bound hydrogen participates in hydrogen bonding with the oxygen of the carbonyl group of a differing amide group. The Figure 4 given in the last section shows that hydrogen bonding plays an analogous role in both the structure of nylon 6,6 and silk.

［注］　arrangement 配置，be responsible for の原因となる，participate in に関与する

和訳例

　　ナイロン中でカルボン酸とアミンを結合している各々のアミド基の窒素上には水素がある。それ（アミド基）はまた，絹の中でアミノ酸由来の単位を結合している。この窒素に結合した水素はすべてのナイロン，すべての種類のタンパク質の全体としての構造に本質的な役割を果している。絹のタンパク質では，この水素はβ-シートの形成に最も顕著に関わっていて，これがタンパク質鎖同士を接触するように保ち，この配列が絹の強さと繊維の性質を生み出している。同じように，ナイロンではこの水素がポリマー鎖間の非常に強い相互作用の原因となり，ナイロンの繊維形成の性質の原因にもなっている。絹でもナイロンでも，窒素に結合した水素が別のアミド基のカルボニル基の酸素との水素結合に関与する。最終項(7)の図(Fig. 4)に示すように，水素結合はナイロン 6,6 と絹の構造両方において同様の役割を演じるのである。

3. The importance of hydrogen bonding can be seen in the properties of water, alcohols, and carboxylic acids among others. A variety of different kinds of covalently bound hydrogen atoms can undergo hydrogen bonding, but the essential necessity for this effect is that the electron deficiency of the hydrogen, covalently bound to an electronegative group, is relieved by forming a weak interaction between this hydrogen and an electron-rich atom. In the amide group the carbonyl oxygen plays the role of the electron-rich atom. In the nylons, many hydrogen bonds can form between different chains. The hydrogen atoms covalently bound to nitrogen are hydrogen-bonded to the carbonyl oxygen of an amide group on a different chain.

［注］　among others いろいろある中で，deficiency 不足，electronegative 電気陰性の，relieve 軽減する，hydrogen bonding 水素結合（すること），hydrogen bond 水素結合，be hydrogen-bonded 水素結合される

［解説］（一部）

The importance of hydrogen bonding can be seen in the properties　of water, alcohols, and carboxylic acids　among others.
水素結合の重要性は，　　　　　　　　　性質に見ることができる。¶水，アルコール，カルボン酸の　¶いろいろある中で
　　　　　　S　　　　　　　　　　Vi（Vt の受身形）

<u>A variety of different kinds of covalently bound hydrogen atoms</u>　<u>can undergo</u>　<u>hydrogen bonding</u>,　but
さまざまな違った種類の　　共有結合した水素原子が　　　　することができる　　水素結合を　　　しかし
　　　　　　　　　　　　　　　　　　S　　　　　　　　　　　　　Vt　　　　　　　O

<u>the essential necessity for this effect</u>　<u>is</u>　<u>that</u>　<u>the electron deficiency of the hydrogen</u>,
この効果の本質的な必要性は，　　　　　　〜にある　〜ということ　　水素の電子不足が
　　　　　S　　　　　　　　　　　　　　　Vi　　　　C (*conj.* 節を率いる)　　(S)

<u>covalently bound to</u>　<u>an electronegative group</u>　<u>is relieved</u>
　⤷ 共有結合した　　　　　⤷ 電気陰性基に　　　　　　　取り除かれる
　　　　　　　　　　　　　　　　　　　　　　　　　　　V (Vtの受動態)

<u>by forming a weak interaction between this hydrogen and an electron-rich atom.</u>
　⤷　この水素と電子過剰な原子の間に弱い相互作用を形成することによって

<u>In the amide group</u>　<u>the carbonyl oxygen</u>　<u>plays</u>　<u>the role of the electron-rich atom</u>
アミド基では，　　　　　　カルボニル酸素が　　　　　演じる　この電子過剰原子の役割を
　　　　　　　　　　　　　　　　S　　　　　　　　　　Vt　　　　O

> [和訳例]
>
> 　水素結合の重要性は，いろいろある中で　水，アルコール，カルボン酸の性質（訳注：分子量のわりに沸点が高いことなど）に見ることができる。さまざまな異なる種類の共有結合した水素原子が水素結合をするが，この効果の本質的な必要性は，電気陰性基（訳注：$N^{\delta-}$）に共有結合した水素の電子不足（訳注：下図参照：$H^{\delta+}$）がこの水素と電子過剰原子（訳注：$O^{\delta-}$）の間の弱い相互作用を形成することによって取り除かれることである。アミド基においては，カルボニル酸素（訳注：$C=O^{\delta-}$）が電子過剰原子の役割を演じる。ナイロンでは，沢山の水素結合が異なる鎖の間に形成し得る。窒素に共有結合した水素原子は別の鎖のアミド基のカルボニル酸素に水素結合される。

訳注：アミド結合の分子間水素結合

4. Hydrogen bonds, which are only about 1/15th as strong as the normal covalent bonds, play a critical role in the properties of nylons. To understand this effect, let's follow the story of nylon 6,6 to the point where the commercially interesting fiber was made in 1939 by Wallace Carothers and Jurian Hill at Du Pont.

　Attractive as nylon 6, 6 seemed, the road to a much-desired serviceable and cheaper replacement for silk stockings was a rocky one. The first problem was that the polyamide fibers lacked tensile strength. Fibers must have very high tensile strength and very low elongation to perform properly. They have to give a little when pulled on, but not too much.

　［注］　attractive 魅力的な，replacement 代替品，tensile strength 引っ張り強さ，
　　　　elongation 伸び
　［解説］（一部）

基礎編 55

```
Attractive as nylon 6,6 seemed,    the road to a much-desired serviceable and cheaper replacement   for silk stockings
ナイロン6,6は魅力的に見えたけれども，非常に渇望されていた，使いやすくて安価な　代替品への道は，　　　絹の靴下の
                                                                        S
was   a rocky one.    The first problem   was that
岩の(険しい)道であった。　　最初の問題は，　　次のこと(that以後の節)であった。
Vi    C             S           Vi    C(conj. 次節を率いる)
the polyamide fibers lacked tensile strength.
このポリアミド繊維が引張強さに欠けていたこと
      (S)       (Vt)        (O)
Fibers must have very high tensile strength and very low elongation    to perform properly.
繊維は，非常に高い引張り強さと 非常に低い伸びを 持たなければならない，　上手く働くためには，
 S    Vt              O                      O
They have to give a little    when pulled on,    but not too much.
少し譲歩し(引っ張られ) ↑引っ張られたときには　　しかし，過度には いけない
なければならない      (引っ張って靴下を履くとき)
  S   Vt   O
```

[和訳例]
　　水素結合は，通常の共有結合の1/15ほどの強さしかないが，ナイロンの性質には決定的な役割を演じる。この効果を理解するために，ナイロン6,6という話を商業的に魅力ある繊維が1939年デュポン社のウォーレス・カローザスとジュリアン・ヒルによって作られた時点までたどることにしよう。
　　ナイロン6,6は魅力的に見えたけれども，非常に渇望されていた，使いやすくて安価な絹の靴下の代替品への道は，岩のような(険しい)道であった。最初の問題は，このポリアミド繊維が引張強さに欠けていたことであった。繊維は，上手く働くには，非常に高い引張強さと 非常に低い伸びを持たねばならない。引っ張られた(引っ張って靴下を履く)ときには少し譲歩し(引っ張られ)なければならないが，過度に譲歩しては(引っ張られては)いけない。

5. Fibers are usually made by extruding polymers, either as melts or as solutions, through a spinneret, a plate containing fine holes. These holes cause the fiber to form as the melt hardens or the solvent evaporates. Melted nylon has a low viscosity because in the melt the hydrogen bonds that hold the polymer molecules together have been broken. Thus, it goes through the spinneret easily. But the fibers made in this way from nylon 6, 6 lacked the strength necessary to be useful.

　［注］　extrude 押し出す，either A or B：AかBかどちらか(どちらでも)，spinneret 紡糸口金，cause to させる

[和訳例]
　　繊維は通常，ポリマーを融解物あるいは溶液として，紡糸口金(細孔を含む板)から，押し出すことによって作られる。これらの孔こそが，融解物が固まるか，あるいは溶媒が蒸発するとともに，繊維の形成をもたらしている。融解したナイロンは低い粘度をもつ：なぜなら，融解状態ではポリマー分子を繋いでいた水素結合は切れてしまっているからである。したがって，それは容易に紡糸口金を通り抜ける。しかし，ナイロン6,6からこのように作られた繊維は，

使うに必要な強度を持たなかった。

6. Jurian Hill melted some nylon in a Petri dish, stirring it with a glass rod. When he withdrew the rod, he observed a filament of the nylon attached to the stirring rod. As he walked with it, it became thinner and, most important, stronger. What happened? The molten nylon molecules originally were in disarray, tangled one with the other. As the nylon was stretched, the polymer molecules lined up, came close together and hydrogen-bonded, the amide groups providing ample opportunity for such bond formation.

[注] Petri dish ペトリ皿（訳注：浅い平底の円形ガラス皿：シャーレ），withdraw 引っこめる，attach くっつく，thinner（thin の比較級）より細く，disarray 乱雑，tangle 絡む，one with the other 互いに，stretch 引っ張る，opportunity 機会

和訳例

ジュリアン・ヒルはいくらかのナイロンをペトリ皿の中で融解させ，ガラス棒で掻き回した。その棒を引き上げたとき，彼はナイロンのフィラメント（細糸）が撹拌棒にくっついてくるのを見た。それ（フィラメント）をもって歩くと，もっと細く，そしてさらに重要なことは，もっと強くなったのである。何が起こったのだろう？ 融解したナイロン分子はもともとは乱雑で，互いに絡み合っていた。ナイロンが引っ張られたとき，ポリマー分子は配列し，互いに接近し，そして水素結合した。アミド基がこのような結合形成に必要な機会を与えたのである。

7. The regular arrangement of the chains enforced by the hydrogen bonding between amide groups along the aligned chains encouraged the regular arrangement of repetitive hydrogen bonds, as seen in the Figure 4 below, and therefore for crystallization to take place. Forming fibers by stretching the melted polymer gives rise to the exceptionally strong material we know nylon to be. The process of stretching is called drawing or orientation and this became an essential step in engineering fiber formation.

[注] enforce 強める，align 並べる，encourage 促す，repetitive 繰り返しの，crystallization 結晶化，exceptionally 例外的に，異常に，drawing 延伸，orientation 配向

和訳例

並んだ鎖に沿ったアミド基間の水素結合によって強められた鎖の規則的な配置は，下図4で見るように，繰り返される水素結合の規則的な配列を促し，それによって結晶化が起こることも促した。融解したポリマーの引き伸ばしによる繊維の形成が，私たちが今ナイロンがそうであることを知っているような，異常に強い材料を与えるのである。引き伸ばしの工程は延伸または配向と呼ばれ，これが工学（強化）繊維形成における必須の一歩となった。

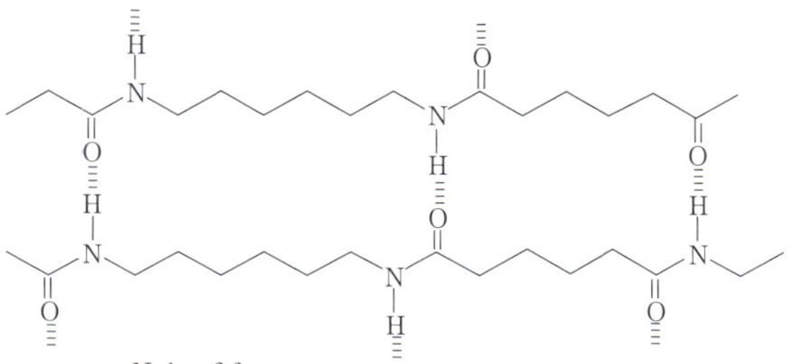

Nylon 6,6

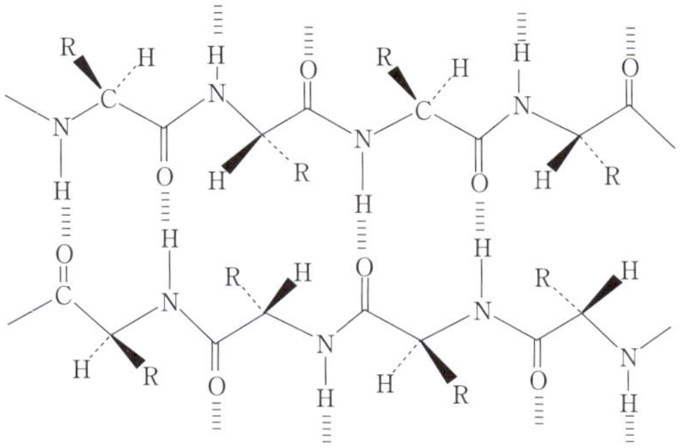

Silk：R=mostly CH₃ (alanine), H (glysine), CH₂OH (serine)

Fig. 4. Schematic hydrogen bonds in nylon 6,6 and silk
（ナイロン 6,6 と絹における水素結合模式図）

[Ex. 24]¹³⁾ Great Art in a Test Tube

1. Organic chemistry has developed into an art form where scientists produce marvelous chemical creations in their test tubes. Mankind benefits from this in the form of medicines, ever-more precise electronics and advanced technological materials. The Nobel Prize in Chemistry 2010 awards one of the most sophisticated tools available to chemists today.

This year's Nobel Prize in Chemistry is awarded to Richard F. Heck, Ei-ichi Negishi and Akira Suzuki for the development of palladium-catalyzed cross coupling. This chemical tool has vastly improved the possibilities for chemists to create sophisticated chemicals, for example carbon-based molecules as complex as those created by nature itself.

Carbon-based (organic) chemistry is the basis of life and is responsible for numerous fascinating natural phenomena: colour in flowers, snake poison and bacteria killing substances such as penicillin. Organic chemistry has allowed man to build on nature's chemistry; making use of carbon's ability to provide a stable skeleton for functional molecules. This has given mankind new medicines and revolutionary materials such as plastics.

In order to create these complex chemicals, chemists need to be able to join carbon atoms together. However, carbon is stable and carbon atoms do not easily react with one another. The first methods used by chemists to bind carbon atoms together were therefore based upon various techniques for rendering carbon more reactive. Such methods worked when creating simple molecules, but when synthesizing more complex molecules chemists ended up with too many unwanted by-products in their test tubes.

Palladium-catalyzed cross coupling solved that problem and provided chemists with a more precise and efficient tool to work with. In the Heck reaction, Negishi reaction and Suzuki reaction, carbon atoms meet on a palladium atom, whereupon their proximity to one another kick-starts the chemical reaction.

Palladium-catalyzed cross coupling is used in research worldwide, as well as in the commercial production of for example pharmaceuticals and molecules used in the electronics industry.

［注］　benefit from から恩恵を受ける，sophisticated 綿密な，cross coupling クロスカップリング：違う有機残基が結合すること：R-X + R'-Y → R-R' (+ XY)

［解説］　本書の初稿を脱稿したとき直後に，Richard F. Heck 博士(1931～；Uiniv. of Delaware 名誉教授)，根岸英一博士(1935～；Perdue Uinv. 教授)，鈴木章博士(1930～；北海道大学名誉教授)3氏の2010ノーベル化学賞受賞の朗報がもたらされた。急遽ノーベル財団のホームページの報道から，1に受賞内容，2に根岸英一氏へのインタビューの一部を引用した。3には鈴木カップリングの概要を引用した。なお，現在は，インタビューや講演の録音・録画もネットで容易に検索することができる。

> 和訳例

試験管内の偉大な芸術

　有機化学は，科学者が試験管の中で膨大な化学の創造品を生み出すという1つの芸術の形に発展している。これによって，人類は医薬，かってない精巧な電子機器，先端技術材料の形で，恩恵を受けている。2010年ノーベル化学賞は，今日化学者が利用できる最も綿密なツールの1つを表彰する。

　本年のノーベル化学賞は，パラジウム触媒クロスカップリングの開発に対して，Richard F. Heck, Ei-ichi Negishi, Akira Suzuki に授与される。この化学のツールは，化学者が綿密な化学品を創造する可能性を広く改良した。例えば自然自体によって創り出されたのと同様に複雑な炭素ベース分子の創造である。

　炭素ベースの(有機)化学は生命の基本であり，無数の魅力的な自然現象を担っている：花の色，蛇毒，ペニシリンのような細菌殺傷物質など。有機化学は，人が自然の化学に基礎をおくことを可能にした；機能性分子の安定な骨格を作り出す炭素の能力を利用して。このことは，人類に新しい医薬や，プラスチックのような革命的な材料を与えた。

　これらの複雑な化学品を作り出すために，化学者は炭素原子を繋ぎ合わせることが必要である。しかし，炭素は安定で，炭素原子はお互いにたやすく反応することはない。したがって，炭素原子を結合するために，化学者の用いた最初の方法は，炭素をもっと反応性にするためのさまざまな技術に基づいていた。このような方法は単純な分子を創るにはうまくいくが，もっと複雑な分子を合成するときは，試験管内に多くの，望まない副生成物を伴ってしまう。

　パラジウム触媒によるクロスカップリングは，この問題を解決し，化学者にもっと精密で，効率的に使える道具を提供した。Heck反応，根岸反応，鈴木反応では，炭素原子はパラジウ

ム原子に出会って，そこで，お互いの近接が化学反応をスタートさせる。
　パラジウム触媒クロスカップリングは世界中の研究で用いられ，また同様に，例えば，調剤薬品や電子産業で使われる分子の商業的生産にも利用されている。

2. Telephone interview with Ei-ichi Negishi following the announcement of the 2010 Nobel Prize in Chemistry, 6 October 2010. The interviewer is Adam Smith, Editor-in-Chief of Nobelprize.org.

　［AS］Was Herbert Brown a very special mentor?

　［EN］Very much so! He really ... In terms of research, he is my only mentor, research mentor. I have had other professors, but he taught me just about everything as to how to do research.

　［AS］What did you learn from him?

　［EN］Well, I must say, true way of doing research.

　［AS］What is the true way of doing research? Can you encapsulate it?

　［EN］Well, so, in many ways, when you pick your subject, or target, or whatever, then we dig out the truth. But, in reality, nobody knows what the truth is. So, we try to do many things to make sure that what we dig out is true. And, many people, in many other cases, people may fall short of that. That will lead to many confusions, of course. And, this search for truth, one finding will lead to another so there's this tremendous scope expanding, you know, in front of you. And, then we continue. So, one of the things that he liked to say is 'a little acorn grow into a tall oak'. And, indeed, that's the mode of our explorations. And, I believe, we some of us, have learned this and how to do it, from him.

　［注］　mentor 助言者，encapsulate 要約する，dig out 調べ上げる，fall short of まで届かない

和訳例

　2010年10月6日，ノーベル化学賞発表に続く，根岸英一博士［EN］との電話インタビュー（一部のみ引用）。
　　インタビュアー：ノーベル賞機構の編集長 Adam Smith ［AS］
［AS］Herbert Brown（訳注：根岸英一氏のポスドク留学先のボス；1979 ノーベル化学賞受賞）は非常に特別な助言者でしたか。
［EN］まさにそうです！彼は本当に…。研究に関しては，彼は私の唯一の助言者，研究の助言者です。他の教授もいましたが，彼がどのように研究をするべきか，まさに全てを教えてくれました。
［AS］彼から何を学びましたか。
［EN］そうですね。研究をする本当の方法と言うべきでしょう。
［AS］研究をする本当の方法といのはどんなことでしょうか。まとめることができますか。
［EN］そうですね，いろんなやり方で，問題，あるいは目標を挙げるとき，あるいは何にしても，私たちは真実を調べ上げます。しかし，現実には，誰も何が真実か知りません。そこで，私たちは多くのことを試みて，調べたことが真実かを確かめます。そして，多くの人は，多くのほかの場合，人々はそこまで届きません。それで，当然，混乱に至りま

す。そして，この真実への探求が，1つの発見が別の発見に導き，その結果，この膨大な展望が，あなたの前に広がっていきますよね。そして，さらに，私たちは続けます。このようにして，彼が好んで言ったことの1つは，「小さいどんぐりは，育って大きい樫の木になる」ということです。そして，これが私たちの探索のし方なのです。そして私は信じています。私たち何人かは，彼から，これを学び，そしてそれをどうするべきかを学んだのです。

3. Suzuki Coupling[14]

$$\text{Ph-B(OH)}_2 + \text{Br-C}_6\text{H}_4\text{-R} \xrightarrow[\text{benzene, }\Delta]{\substack{\text{2eq K}_2\text{CO}_3\text{aq.}\\ \text{3mol-\%Pd(PPh}_3)_4}} \text{Ph-C}_6\text{H}_4\text{-R}$$

The scheme above shows the first published Suzuki coupling, which is the palladium-catalysed cross coupling between organoboronic acid and halides. Recent catalyst and methods developments have broadened the possible applications enormously, so that the scope of the reaction partners is not restricted to aryls, but includes alkyls, alkenyls and alkynyls. Potassium trifluoroborates and organoboranes or boronate esters may be used in place of boronic acids. Some pseudohalides (for example triflates) may also be used as coupling partners.

> 和訳例

鈴木カップリング

　上の図は，最初に公表された鈴木カップリングを示す。これはパラジウム触媒による有機ボロン酸とハロゲン化物間のクロスカップリングである。最近の触媒および手法の発展によって可能な応用は大きく広がったので，反応相手の範囲は，アリール（芳香族類）に限らず，アルキル（脂肪族），アルケニル（含二重結合），アルキニル類（含三重結合）を含む。ボロン酸類の代わりに，トリフルオロホウ酸カリウム（K BF$_3$-R）および有機ボランあるいはボロン酸エステルが用いられる。いくつかの偽ハロゲン化物（例えばトリフレート）もカップリングの相手として用いられる。
　　　　　　　　　　　　　　　訳注：ボロン酸 = R-B(OH)$_2$ (R = alkyl, aryl)

演習編

Only your own efforts will polish your skill in English and in chemistry!
(君自身の努力だけが君の英語と化学の能力を磨く)

[Ex. 25]¹⁾ Chemistry and Society

1. Chemistry is the science that examines the molecular reasons for macroscopic phenomena. It uses the scientific method, which emphasizes observation and experiment, to explore the link between our everyday world and the world of molecules and atoms.

　［注］　macroscopic 巨視的，(*cf.* microscopic 微視的), phenomena (phenomenon の複数) 現象，emphasize 強調する，explore 探求する

2. As early as 600 B.C., people wondered about the underlying reasons for the world and its behavior. Several Greek philosophers, including Plato, Democritus, Thales, Empedocles, and Aristotle, believed that reason alone could unravel the mysteries of nature. They made some progress in our understanding of the natural world and introduced fundamental ideas such as atoms and elements. Alchemy, the predecessor of chemistry, flourished in the Middle Ages and contributed to chemical knowledge, but because of its secretive nature, knowledge was not efficiently propagated and progress came slowly.

　［注］　as early as ほども早くに，underlying (underlie (の下にある) の現在分詞) 根本的な，that (目的語節を率いる接続詞) 〜ということを，unravel 解決する，fundamental 基本的な，alchemy 錬金術，predecessor 先行者，flourish 栄える，contribute 寄与する，secretive 秘密的な，propagate 伝播する

3. In the sixteenth century, scientists focused on observation as the key to understanding the natural world. Books authored by Copernicus and Vesalius exemplify this change of perspective and mark the beginning of the scientific revolution. These books were followed by relatively rapid developments in our understanding of the chemical world. Boyle initiated a scheme that can be used to classify matter according to composition. Lavoisier and Proust formulated the laws of conservation of mass and constant composition, respectively. John Dalton built on these laws to develop the atomic theory. Rutherford then examined the internal structure of the atom and found it to be mostly empty space with a very dense

nucleus in the center and negatively charged electrons orbiting around it. The foundations of modern chemistry were laid.

[注] exemplify 例証する，perspective 考え，revolution 革命，respectively それぞれ，be followed by に引き継がれる，build on に頼る，orbit 回る

4. Understanding chemistry deepens our understanding of the world and our understanding of ourselves because all matter, even our brains and bodies, are made of atoms and molecules.

[注] be made of からできている

5. The Greek way of thinking impacted society, or rather, delayed the impact that science would have. Few questioned this method and as a result, science and technology were largely ignored for hundreds of years. If you doubt this, consider how our world might be if the scientific revolution happened among the Greeks in 600 A.D. Our society has profoundly changed after only 450 years of scientific progress. Where would it be after 2600 years?

[注] impact(vt)影響を与える，(n)影響，ignore 無視する

6. The social changes that resulted from shift in viewpoint were major. Science began to flourish as we began to learn how to understand and control the physical world. Technology grew as we began to apply the knowledge acquired through the scientific method to improving human lives. If doubt this, think about your own life — how would it be different without technology? Along with those changes came the responsibility to use knowledge and technology wisely — only society as a whole can do that. The power that science bestowed on us has been used to improve society — think about advances in medicine as an example — but it has also been used to destroy — think of the atomic bomb or pollution. How do we, as a society, harness the power that science has given for good and avoid the bad?

[注] acquire 獲得する，responsibility 責任，bestow 授ける，harness 利用する

[Ex. 26][15] SI System

Nearly all the world's people use a set of measurement units called Le Système International deUnités, commonly referred to as the SI system. The SI system is a recent modification of the older metric system. The SI system is popular because it is a decimal number system. Every unit is ten times of the next smaller unit, for example there are 10 millimeters in 1 centimeter. The familiar English system of measurement is not a decimal system — there are 4 quarts in 1 gallon, 12 inches in 1 foot, and 16 ounces in 1 pound.

The SI system uses a small number of base units (such as the meter) and a set of prefixes (such as kilo-, centi-, and milli-) which can be used with any base unit.

Chemists continue to use a few older metric units such as the liter and the milliliter that are not part of the SI system. A volume of 1 liter (1 L) equals 1 cubic decimeter (1 dm^3) and 1 milliliter (1 mL) is equal to 1 cubic centimeter (1 cm^3).

Measured quantities almost always consist of two parts : a number and a unit. You must get in the habit of writing units with your measured numbers. Don't write a length as 5 ; write it as 5 mm, 5 cm, or 5 m.

［注］　Le Système International deUnités（フランス語）= International system of units（SI system）国際単位系（SI 系），modification 変形，修飾，decimal 十進の，prefix 接頭語，equal = be equal to に等しい，consist of から成る。

[Ex. 27][1] Periodic Law

1. In the 1860s a popular Russian professor named Dmitri Mendeleev (1834-1907) at the Technological Institute of St. Petersburg wrote a chemistry textbook. He drew on the growing knowledge of the descriptive chemistry and noticed that elements could be grouped into families having similar properties. Some elements, such as helium, neon, and argon, were all chemically inert gases; they did not react to form compounds. Others, like sodium and potassium, were reactive metals.

［注］　descriptive 記述的な，group into 分類する，inert 不活性な，compound 化合物

2. Furthermore, he found that if he listed the elements in order of increasing atomic weight, these similar properties would recur in a periodic fashion. Mendeleev summarized these observations in the periodic law that states, "When the elements are arranged in order of increasing atomic weight, certain sets of properties recur periodically."

［注］　furthermore さらに，in order of～：～の順に，recur 繰り返す，periodic 周期的な，periodic law 周期律

3. Mendeleev then organized all the known elements in a table so that atomic weight increased from left to right and elements with similar properties aligned in the same vertical columns. For this to work, Mendeleev had to leave gaps in his table. He predicted that elements would be discovered to fill these gaps. He also had to propose that some measured atomic weights were erroneous. In both cases, Mendeleev was correct.

［注］　organize 組織する，整理する，so that～：～するように，align 並ぶ，vertical 縦の，predict 予測する，propose 提案する

4. Within 20 years of Mendeleev's proposal, three gaps were filled with the discoveries of gallium (Ga), scandium (Sc), and germanium (Ge). Mendeleev's arrangement of elements is called the periodic table and is foundational to modern chemistry.

［注］　proposal 提案，foundational 基本の

5. Mendeleev did not know why the periodic law existed. His law, like all scientific laws, summarized a large number of observations but did not give the underlying reasons for the observed behavior. The next step in the scientific method was to devise a theory that explained the law and provided a model for atoms. In Mendeleev's time, there was no theory to explain the periodic law.

［注］　underlying 下にある，基礎となる

6. We take a brief look at a theory that explains why the properties of elements recur in a periodic fashion. The theory is called the Bohr model for the atom, after Niels Bohr (1885-1962), and it links the macroscopic observation — that certain elements have similar properties that recur — to the microscopic reason — that the atoms comprising the elements have similarities that recur.

［注］　take a brief look at ～：～を少し見る，macroscopic 巨視的な，microscopic 微視的な，link A to B：A を B に結びつける

[Ex. 28][1] Radioactivity

1. Radioactivity, discovered by Becquerel and the Curies, consists of energetic particles emitted by unstable nuclei. Alpha radiation consists of helium nuclei that have high ionizing power but low penetrating power. Beta radiation consists of electrons emitted when a neutron within an atomic nucleus converts into a proton. Beta particles have lower ionizing power than alpha particles but higher penetrating power. Gamma radiation is high-energy electromagnetic radiation with low ionizing power but high penetrating power. Unstable nuclei radioactively decay according to their half-life, the time it takes for one-half of the nuclei in a given sample to decay.

［注］　radioactivity 放射能，the Curies キュリー夫妻(Marie Curie and Pierre Currie)，consist of ～　～から成る，emit 発する，nuclei（複）nucleus（単）核（原子核），ionizing power イオン化力，penetrating power 透過（貫通）力，
convert into ～ = convert to ～　～に変換する，electromagnetic radiation 電磁波放射線（電磁線），according to ～：～に従って，～によれば，decay(vi)崩壊する

2. Some heavy elements such as U-235 and Pu-239 can become unstable and undergo fission when bombarded with neutrons. The atom splits to form lighter elements, neutrons, and energy. If fission is kept under control, the emitted energy can be used to generate electricity. If fission is forced to escalate, it results in an atomic bomb. Hydrogen bombs, similar to the sun, employ a different type of nuclear reaction called fusion in which the nuclei of lighter elements combine to heavier ones. In all nuclear reactions that produce energy, some mass is converted to energy

in the reaction.

[注] fission (核)分裂, bombard 衝撃(爆撃)する, split 分裂する, generate 発生する, force to ～(強制的に)～させる, result in ～(結果として)～を生じる, fusion 融合

3. By measuring the levels of certain radioactive elements in fossils or rocks, radioactivity can be used to date objects. The age of earth is estimated to be 4.5 billion years based on the ratio of uranium to lead in the oldest rocks. High levels of radioactivity can kill human life. Lower levels can be used in therapeutic fashion to either diagnose or treat disease.

[注] certain ある(種の), fossil 化石, date 日付をする, 年代を推定する, ratrio of A to B A 対 B 比(A/B), therapeutic 治療の, diagnose (vt)診断する

4. The discovery of radiation has had many impacts on our society. It ultimately led to the Manhattan Project, the construction and detonation of the first atomic bomb in 1945. For the first time, in a very tangible way, society could see the effects of the power that science had given to it. Yet science itself did not drop the bomb on Japan. It was the people of the United States who did that, and the question remains — how do we use the power that technology can give? Since then, our society has struggled with the ethical implications of certain scientific discoveries. For the past decade, nuclear weapons have been disarmed at the rate of 2000 bombs per year. Today, we live in an age when the threat of nuclear annihilation is less severe.

[注] impact 影響, ultimately 究極的に, detonation 爆発, tangible 実体的な, 明確な, society 社会の人々, struggle with 苦闘する, ethical 倫理的な, implication 意味, 連携, disarm 軍備を解く, 軍備縮小する, annihilation 絶滅

5. Nuclear fission is used to generate electricity without harmful side effects associated with fossil fuel combustion. Yet nuclear power has its own problems, namely the potential for accidents and waste disposal. Will the United States build a permanent site for nuclear waste disposal? Will we turn to nuclear power as the fossil fuel supply dwindles away? How many resources will we put into the development of fusion as a future energy source? These are all questions that our society faces as we begin this new millennium.

[注] waste disposal 廃棄物処理, dwindle away (vi)減少してなくなる, millennium 一千年

6. Nuclear processes have been able to tell us how old we are. Archaeological discoveries are fitted into a chronological puzzle that tells about human history from the very earliest times. We know that billions of years passed on the earth before humans ever existed. We know how certain humans began to use tools and how they migrated and moved around on the earth. We can date specific items such as the Shroud of Turin and determine if they are genuine. What effect does the scientific

viewpoint have on our society? On religion? What does it tell us about who we are?

[注] archaeological 考古学的, chronological 年代順の, the Shroud of Turin トリノの経かたびら(キリストの着衣), genuine 真の

[Ex. 29]⁷⁾ Carbon Isotopes

1. Natural carbon contains three isotopes: carbon-12 (98.89%), the most prevalent isotope; a small proportion of carbon-13 (1.11%); and a trace of carbon-14. Carbon-14 is a radioactive isotope with a half-life of 5.7×10^3 years. With such a short half-life, we would expect little sign of this isotope on earth. Yet it is prevalent in all living tissue, because the isotope is constantly being produced by reactions between cosmic ray neutrons and nitrogen atoms in the upper atmosphere:

$${}^{14}_{7}\text{N} + {}^{1}_{0}\text{n} \longrightarrow {}^{14}_{6}\text{C} + {}^{1}_{1}\text{H}$$

[注] prevalent 広く行き渡っている, half-life 半減期, upper atmosphere 高層大気(対流圏より上方)

2. The carbon atoms react with oxygen gas to form radioactive molecules of carbon dioxide. These are absorbed by plants in photosynthesis. Creatures that eat plants and creatures that eat the creatures that eat plants will all contain the same proportion of radioactive carbon, and the carbon-14 already present in the body decays. Thus, the age of an object can be determined by measuring the carbon-14 present in a sample. This method provides an absolute scale of dating objects that are between 1,000 and 20,000 years old. W. F. Libby was awarded the Nobel Prize in chemistry in 1940 for developing the radiocarbon-dating technique.

[注] photosynthesis 光合成, decay 崩壊する
that いずれも(接続詞＋主語)の役割をする関係代名詞
dating 年代を算定する (初めの -ing は動名詞, 後の -ing は現在分詞)

[Ex. 30]¹⁵⁾ Carbon Dioxide, Henry's Law, and Absolute Zero.

1 Carbon dioxide gas is one of the end products of digesting food and it is in the air we exhale, but our exhaled breath is only 4% CO_2. The atmosphere we breathe contains 0.03% CO_2. The presence of carbon dioxide in the blood stimulates breathing. For this reason, carbon dioxide is added to oxygen in artificial respiration and to the gases used in anesthesia.

[注] digest 消化する, exhale (息を)吐き出す cf. inhale (息を)吸入する, breathe (v)呼吸する, breath (n)息, stimulate 刺激する, respiration 呼吸, cf. respire (v)呼吸する, anesthesia 麻酔

2. Carbon dioxide is also produced when any organic (carbon-containing) fuel burns. Fuels such as gasoline, coal, natural gas, wood and paper all produce CO_2

when they burn. Burning too many of these fuels may add too much carbon dioxide to our atmosphere — that can lead to overheating of the earth in a process known as global warming. Plants remove carbon dioxide from the air and make food from it in the photosynthesis process. Carbon dioxide is also removed from the atmosphere by dissolving in seawater.

　［注］　global warming 地球温暖化，　burning を燃やすこと（動名詞，主語）

3. Carbonated beverages such as soda contain dissolved carbon dioxide. In baking cookies or cakes, a chemical reaction produces CO_2 from either baking soda or baking powder. The carbon dioxide gas produces the tiny holes in cookies and cakes. In baking bread or pizza dough, yeast produces carbon dioxide from sugar — the CO_2 makes the dough rise.

　［注］　carbonated beverage 炭酸飲料，baking soda 重曹（$NaHCO_3$），baking powder ふくらし粉，make the dough rise パン生地をふくらませる

4. Carbon dioxide gas can extinguish a fire. It does this by reducing the amount of oxygen in the air around the burning object. Oxygen in the air is necessary for a fire and when this oxygen is removed, the fire goes out.

　［注］　extinguish a fire 火を消す，go out 消える

5. Although carbon dioxide is a gas at room temperature, it is a solid at temperature below $-78.5℃$. Solid carbon dioxide is called "dry ice" since it changes from solid to gas without ever forming a liquid, a process called sublimation.

　［注］　sublimation 昇華

6. [Henry's law] According to Henry's law, the solubility of a gas in a liquid is directly proportional to the pressure of the gas on the liquid. Increasing the pressure increases gas solubility; decreasing the pressure decreases its solubility. An unopened bottle of a carbonated beverage contains a large amount of carbon dioxide gas dissolved in the solution. While dissolved, the gas is invisible — the mixture is homogeneous. An open bottle of carbonated beverage has bubbles of carbon dioxide gas rising to the surface. These bubbles of gas are not dissolved. Opening the bottle decreases the pressure and decreases the solubility of carbon dioxide in water.

　［注］　directly proportional 正比例の，invisible 目に見えない，homogeneous 均一の

7. [Absolute Zero] Temperature is a measure of the average kinetic energy of the molecules in a sample of a matter. At room temperature, molecules are moving very

rapidly. When molecules slow down, their temperature decreases. Theoretically, at some very cold temperature the motion of the molecules would slow to zero and the molecules would be totally motionless. The temperature, called absolute zero, is equal to $-273.15°C$ or 0 K. Scientists have recently achieved a temperature of 0.000000002 K, very close to absolute zero. Modern theory says that absolute zero can never be reached; molecular motion can never completely cease.

[注] absolute zero 絶対零度, kinetic energy 運動エネルギー, slow down 遅くなる, motionless 動かない

8. In a rigid container (one in which the volume remains constant), increasing the temperature will increase the pressure of the gas. Lowering the temperature will lower the pressure. This relationship is sometimes known as the Pressure-Temperature Law, which states that the pressure of a gas varies with the absolute temperature provided that the volume remains constant. For an experiment, air is filled in a hollow steel ball attached with a pressure gauge. Immerse the ball in liquids at different temperatures and measure the pressure in the ball. Plot the data of temperature and pressure on a graph paper, and draw a best-fitting straight line. Extrapolate the line. Absolute zero is the temperature at which the pressure would be zero.

[注] rigid container 硬い容器, provided that ～：～の条件で, attached with a pressure gauge 圧力計を付した, extrapolate 外挿する

[Ex. 31][1] Ionic and Covalent Compounds

1. Most of the substances in nature are compounds, combinations of elements in fixed ratios. Compounds are represented by their chemical formulas, which at a minimum identify the type and relative amounts of each element present. For covalent compounds, a more specific kind of chemical formula, called a molecular formula, indicates the number and type of atoms in each molecule.

[注] compound 化合物, combination 組合せ, represent(vt)表す, at a minimum 最低限, identify(vt)同定する, ～ present 存在する～（通常，先行する名詞を形容する）, covalent 共有結合の, specific 明確な

2. Chemical compounds are divided into two types, ionic and covalent, each with their own naming system. An ionic compound is a metal bonded to a nonmetal via an ionic bond. In an ionic bond, an electron is transferred from the metal to the nonmetal, making the metal a cation (positively charged) and the nonmetal an anion (negatively charged). In its solid form, an ionic compound consists of a three-dimensional lattice of alternating positive and negative ions. A covalent compound is a nonmetal bonded to a nonmetal via a covalent bond. In a covalent bond, electrons

are shared between the two atoms. Covalent compounds contain identifiable clusters of atoms called molecules. The properties of molecules determine the properties of the covalent compound they compose.

[注] divide into 〜：〜に分類する，naming 名前を付ける(命名の)，cation カチオン(陽または正イオン)，anion アニオン(陰または負イオン)，lattice 格子，alternating 交互の，share 共有する，identifiable 同定できる，cluster かたまり，compose 構成する，consist of 〜：〜から成る

3. The sum of the atomic weights of all the atoms in a molecular formula is called the molecular weight. It is a conversion factor between mass of the compound and moles of its molecules. Chemical reactions, in which compounds are formed or transformed, are represented by a chemical equation. The substances on the left side of a chemical equation are called the reactants, and the substances on the right side are called the products. The number of atoms of each type on each side of the chemical equation must be equal in order for the equation to be balanced. The coefficients in the chemical equation can help us determine numerical relationships between the amounts of reactants and products.

[注] conversion factor 変換因子，transform 変化する，reactant 反応物，product 生成物，coefficient 係数，numerical 数の，in order for ― to〜：―が〜するため(よう)に，help ―(to)〜：―が〜するのを助ける，―に〜させる

4. Because most of the substances around us are compounds, we must understand them to understand what is happening around us. Compounds, or the molecules that compose compounds, are important in everything from the materials we use every day ― such as plastics, detergents, or antiperspirants ― to the environmental problems that we face as a society ― such as ozone depletion, air pollution, or global warming.

[注] detergent 洗剤，antiperspirant 発汗抑制剤，environment 環境，depletion 枯渇

5. Ionic compounds are found in food ― table salt, for example, is NaCl ― and in seawater and soils. Covalent compounds compose water, most of the fuels we burn, much of the food we eat, and most of the molecules important in life.

6. It is not an understatement to say that chemical reactions keep both our society and our own bodies going. Ninety percent of our energy is derived from chemical reactions, primarily the combustion of fossil fuels. Our own bodies derive energy from the foods we eat by orchestrating a slow combustion of the molecules contained in food.

[注] understatement 控えめな表現 cf. overstatement 誇張表現，keep〜going：〜が活動を

続ける，derive A from B：A を B から引き出す，A を B に由来する，primarily 主として，combustion 燃焼

7. The products of useful chemical reactions can sometimes present environmental problems. For example, carbon dioxide, one of the products of fossil-fuel combustion, may be causing the planet to warm through a process called the greenhouse effect.

　［注］　cause ―(to)～：―が～するのを引き起こす，―に～させる，greenhouse 温室

[Ex. 32]¹⁶⁾ Energy and Working

When a force acts on an object and moves the object, the change in the object's kinetic energy is equal to the work done on the object. Work has to be done, for example, to accelerate a car from 0 to 60 miles per hour or to hit a baseball out of a stadium. Work is also required to increase the potential energy of an object. Thus, work has to be done to raise an object against the force of gravity (as in an elevator), to separate a sodium ion (Na^+) from a chloride ion (Cl^-), or to move an electron away from an atomic nucleus. The work done on an object corresponds to the quantity of energy transferred to the object ; that is, doing work is a process that transfers energy to an object. Conversely, if an object does work on something else, the quantity of energy associated with the object must decrease.

　注：kinetic energy 運動エネルギー，correspond to ～：～に相当する，対応する

[Ex. 33]⁶⁾ Introduction to Chemical Thermodynamics

1. An alert young scientist with only an elementary background in his/her field might be surprised to learn that a subject called "thermodynamics" has any relevance to chemistry, biology, material science, and geology. The term *thermodynamics*, taken literally, implies a field concerned with the mechanical action produced by heat.

　［注］　alert 油断のない，利発的な，geology 地質学，relevance 関係，literally 文字通りに，imply 意味する

2. Lord Kelvin (William Thomson) invented the name to direct attention to the *dynamic* nature of *heat* and to contrast this perspective with previous conceptions of heat as a type of fluid. The name has remained, although the applications of the science are much broader than when Kelvin created its name.

　［注］　invent 発明する，perspective 見通し，conception 概念，remain 残る

3. In contrast to mechanics, electromagnetic field theory, or relativity, where the names of Newton, Maxwell, and Einstein stand out uniquely, the foundations of thermodynamics arose from the thinking of over one half a dozen individuals: Carnot, Mayer, Joule, Helmholtz, Kelvin, Clausius. Each provided critical steps leading to the grand synthesis of the two classical laws of thermodynamics.

［注］　in contrast to (with) ～　～と対照的に，relativity 相対論，関連性，stand out 抜きん出る

[Ex. 34] [17] Objectives of Chemical Thermodynamics

1. In practice, the primary objective of chemical thermodynamics is to establish a criterion for determining the feasibility or spontaneity of a given physical or chemical transformation. For example, we may be interested in a criterion for determining the feasibility of a spontaneous transformation from one phase to another, such as the conversion of graphite to diamond, or the spontaneous direction of a metabolic reaction that occurs in a cell. On the basis of the first and second laws of thermodynamics, expressed in terms of Gibbs energy functions, several additional theoretical concepts and mathematical functions have been developed that provide a powerful approach to the solution of these questions.

［注］　（次のEx. 35とともに，ルイスの熱力学の古典的名著から。）
　　　in practice 実際上，objective 目的，criterion 規準，feasibility（実行の）可能性，spontaneity 自発性，metabolic 代謝の，solution 解，解決

2. Once the spontaneous direction of a nature process is determined, we may wish to know how far the process will proceed before reaching equilibrium. For example, we might want to find the maximum yield of an industrial process, or the equilibrium solubility of atmospheric carbon dioxide (CO_2) in natural waters, or the equilibrium concentration of a group of metabolites in a cell. Thermodynamic methods provide the mathematical relations required to estimate such quantities.

［注］　once ～：いったん～すると，how far ～：どこまで～か，atmospheric 大気の，metabolite 代謝物質

3. Although the main objective of chemical thermodynamics is the analysis of spontaneity and equilibrium, the methods are also applicable to many other problems. For example, the study of phase equilibria, in ideal and nonideal systems, is basic to the intelligent use of the techniques of extraction, distillation, and crystallization, to metallurgical operations, to the development of new materials, and to the understanding of the species of minerals found in geological systems. Similarly, the energy changes that accompany a physical or chemical transformation, in the form of either heat or work, are of great interest, whether the transformation is the combustion of a fuel, the fission of a uranium nucleus, or the transport of a

metabolite against a concentration gradient. Thermodynamic concepts and methods provide a powerful approach to the understanding of such problems.

[注] equilibria(equilibrium 平衡)の複数, extraction 抽出, metallurgical 冶金の, against～：～に反する(逆らった), ～に対する

[Ex. 35][17] The First, Second, and Third Laws of Thermodynamics

1. Clausius summed up the findings of thermodynamics in the statement, "Die Energie der Welt ist constant; die Entropie der Welt strebt einem Maximum zu", and it was this quotation which headed the great memoir of Gibbs on "The Equilibrium of Heterogeneous Substances". What is this entropy, which such masters have placed in a position of coordinate importance with energy, but which has proved a bugbear to so many a student of thermodynamics?

[注] "Die Energie der Welt ist constant ; die Entropie der Welt strebt einem Maximum zu"（ドイツ語）："The energy of the world is constant ; the entropy of the world strives to a maximum." head 頭となる, 率いる, coordinate 同格の, bugbear 怖いもの, お化け

2. The first law of thermodynamics, or the law of conservation of energy, was universally accepted almost as soon as it was stated, not because the experimental evidence in its favor was at that time overwhelming, but rather because it appeared reasonable and in accord with human intuition. The concept of the permanence of things is one which is possessed by all. It has been extended from the material to the spiritual world. The idea that, even if objects are destroyed, their substance is in some way preserved has been handed down to us by the ancients, and in modern science the utility of such a mode of thought has been fully appreciated. The recognition of the conservation of carbon permits us to follow, at least in thought, the course of the element when coal is burned and the resulting carbon dioxide is absorbed by living plants, whence the carbon passes through an unending series of complex transformations.

[注] conservation 保存, as soon as や否や, overwhelming 圧倒的な, appear に見える, 思われる, in accord with と一致して, intuition 直感, preserve 保存する, recognition 認識, permit させる, resulting 生じた, whence そこから, not ―, but (rather)～：―でなくて, (むしろ)～

3. The second law of thermodynamics, which is known also as the law of dissipation or degradation of energy, or the law of the increase of entropy, was developed almost simultaneously with the first law through the fundamental work of Carnot, Clausius, and Kelvin. But it met with a different fate, for it seemed in no recognizable way to accord with existing thought and prejudice. The various laws of

conservation had been foreshadowed long before their acceptance into the body of scientific thought. The second law came as a new thing, alien to traditional thought, with far-reaching implications in general cosmology.

　［注］　dissipation 散逸，degradation 退化，prejudice 先入観，偏見，foreshadow 予示する，alien 相容れない，外国の，異なる，far-reaching 遠くに及ぶ，広範囲の，cosmology 宇宙論

4. Because the second law seemed alien to the intuition, and even abhorrent to the philosophy of the time, many attempts were made to find exceptions to this law and thus to disprove its universal validity. But such attempts have served rather to convince the incredulous and to establish the second law of thermodynamics as one of the foundations of modern science. In this process we have become reconciled to its philosophical implications or have learned to interpret them to our satisfaction; we have learned its limitations, or, better, we have learned to state the law in such a form that these limitations appear no longer to exist; and especially we have learned its correlation with other familiar concepts, so that now it no longer stands as a thing apart, but rather as a natural consequence of long-familiar ideas.

　［注］　abhorrent 相容れない，convince 納得させる，incredulous 懐疑的な，the incredulous 懐疑派，reconcile 和解させる，調和させる，long-familiar 長く親しんできた

5. During the 19th century the first and second laws of thermodynamics were formulated and applied to many scientific problems. It was not until the first decade of the 20th century that investigations at very low temperatures had progressed sufficiently to lay the basis for the generalization which constitutes the third law of thermodynamics. Nevertheless, the third law has now satisfied numerous and searching experimental tests and has provided the basis for a large fraction of data on chemical reactions. It seems appropriate, therefore, to include the third law as an integral portion of the basic foundation of our subject.

　［注］　formulate 定式化する，generalization 一般化，一般則，integral 欠くことのできない，searching 厳重な，it seems～to—：—するのが～と思われる

6. We have now seen that heat capacities of solids decrease very rapidly to zero as the absolute zero of temperature is approached. Also we recall the explanation of Einstein that the internal energy of a solid is quantized so that only finite quanta of energy can be absorbed. Thus, as the solid is cooled to 0 K, all its constituent particles fall into their lowest quantum states; indeed we may say that the entire crystal is in its lowest quantum state. This is evidently an especially simple state, which we may take as an absolute unit of probability.

［注］　heat capacity 熱容量，quantize 量子化する，quanta（複数），quantum（単数）量子，constituent 構成成分，probability 確率

[Ex. 36] [18) Number and Choice of Components

1. Gibbs wrote: "If the conditions are satisfied, the choice of the substances which we are to regard as the components of the homogeneous mass considered, may be determined entirely by convenience, and independently of any theory in regard to the internal constitution of the mass." The conditions involve the number p of the components, which must be such that the values of the differentials $dn_1, dn_2, \cdots, dn_p$ "shall be independent, and shall express every possible variation in the composition of the homogeneous mass considered…," In accordance with this principle it is possible to define a potential μ_i for any substance known or imagined to be present in a phase under consideration, but some of these μ_i values may turn out to be identical, and some may be interrelated in various ways.

［注］　condition 条件，component 成分，homogeneous 均一な，convenience 便利，都合，differentials 微分，independently of～：～に無関係に，in accordance with～：～と一致して，turn out to～：～であることが分かる

2. With respect to the number of components the time scale of the observation is important. At room temperature and in the absence of a catalyst a mixture of hydrogen, oxygen, and water is a three-component system, because there is no interconversion within a humanly possible time of observation. At high temperature or in the presence of a catalyst it is a two-component system, because the interconversion is so fast that over any humanly possible time of observation the proportion of water is determined by the ultimate composition of the system. Under intermediate conditions the behavior of the system will approach that of a three-component system for rapid observations and that of a two-component system for observations made on a slow time scale.

［注］　interconversion 相互変換，intermediate 中間の

[Ex. 37] [18) The Tubular-Flow Reactor

1. The twin problems of mixing of reactants and measurement of time interval present no serious difficulties in the study of a reaction with a half time of an hour or more, i.e., one in which the time required for the reaction of half of the reactant initially present is as much as an hour. They present difficulties (but not insuperable ones) in the way of the precise estimation of specific rate for reactions with half times in the interval from an hour to a second. For faster reactions, the time problem

can be alleviated by estimating the extent of reaction by way of an automatic device which records as a function of time the value of a physical property which adjusts rapidly to the composition of the solution. It is a considerable advantage if the value of the property is linearly related to the extent of reaction.

　［注］　twin 双子の，対の，half time 半減期，insuperable 克服できない，alleviate 軽減する

2. An effective way of exploiting the advantages of this kind of measurement and at the same time of alleviating the mixing problem is the tubular-flow reactor. The development of this device by Hartridge and Roughton extends the limits of usefully measurable reaction rates to systems with half time of as little as 0.001 sec and opened up a region of reaction kinetics which had previously been completely inaccessible. Techniques of operations have been highly developed, much has been done in the way of automation, and the results have been important.

　［注］　tubular-flow reactor 管流動反応器，extend to 〜：〜に拡張する

3.　In this technique, two solutions of liquid flowing at constant velocity are rapidly and thoroughly mixed and allowed to pass through a tube of constant diameter. Under favorable conditions the composition of a thin cylindrical lamina of solution at a distance l from the point of mixing is then the same as if the mixed solution had reacted without flow for a time interval t given by : $t = Al/u$, where A is the area of cross section of the tube and u is the flow rate in volume per unit time. The measurement of a physical property which is dependent on extent of reaction as a function of the distance l supplies therefore the same information about the kinetics of the reaction as measurement of the property as a function of time would in a nonflowing system.

　［注］　allow to〜: 〜させる，lamina 薄層，cross section 断面，the same〜as —：—と同じ〜

4. It is critically important in this kind of measurement that the composition of the solution be essentially constant throughout any perpendicular cross section of the tube, that mixing and diffusion parallel to the axis of the tube be negligible, and especially that observations should begin at a point at which mixing of the input solution is essentially complete. In an incompletely mixed system, reaction occurs in a hodgepodge of volume elements of varying composition, and the observed extent of reaction is meaningless.

　［注］　critically 決定的に，perpendicular 垂直の，直交する，diffusion 拡散，parallel to 〜：〜に平行な，hodgepodge ごた混ぜ，it — that〜：it は that〜の仮主語(形式主語)：〜(であることは)—(である)，—(なのは)〜(ということである)

[Ex. 38][19] Soap as an Amphiphilic Compound

A surface active molecule consists of two parts with opposing character — like Dr. Jekyll and Mr. Hyde. One part is hydrophilic and the other is hydrophobic. A typical surface active molecule, sodium octadecanoate (one of the main components of soap), is shown below. As shown in the figure, surface active molecules are usually drawn like a match, with the head of a match being a hydrophilic group (which is the sodium carboxylate group in the case of soap) and the stem being the hydrophobic group. The hydrophobic group is usually a hydrocarbon chain, and so it is often called a lipophilic group. The substance that shows surface activity is called by several names — surface active substance, surface active compounds, and surface active agent (surfactant). Since surface active substances exhibit affinity to both water and oils, they may also be called amphiphilic compounds, amphiphiles, or amphoteric compounds.

Hydrophobic or lipophilic stem　　*Hydrophilic head*

［注］ hydrophilic 親水性の，hydrophobic 疎水性の，lipophilic 親油性の，amphiphilic 両親媒性の，amphiphile 両親媒性種，amphoteric 両性の

[Ex. 39][19] Origin of Surface Tension

Condensed matters (liquids and solids) have surface tension because cohesive energy is present between their molecules. A molecule in the bulk of a condensed matter interacts with attractive force with its surrounding molecules. For example, a water molecule in bulk liquid phase makes some (at most four) hydrogen bonds as well as van der Waals interactions, and a carbon atom in a diamond crystal has four C–C covalent bonds with nearest neighbor atoms. Molecules present at a surface, however, cannot fully form such bonding and/or interaction since they have no (or few) interacting molecule in the vacuum (vapor) side and thus have excess energy compared with those in the bulk phase. This excess energy existing in the surface molecules or atoms is defined as surface tension.

［注］ surface tension 表面張力，cohesive energy 凝集エネルギー，bulk 本体，内部，compared with と比較して

[Ex. 40][10] Electromagnetic Radiation

1. Common to most of the characterization techniques for organic compounds is electromagnetic radiation. Electromagnetic radiation is light: not only the light we see, but the light we cannot see as well. Even though the electromagnetic spectrum represents a continuous range of energy, it can be divided into regions such as γ

-rays, X-rays, ultraviolet, visible, infrared, microwaves, and radio waves, in decreasing order of energy, or in increasing order of wavelength.

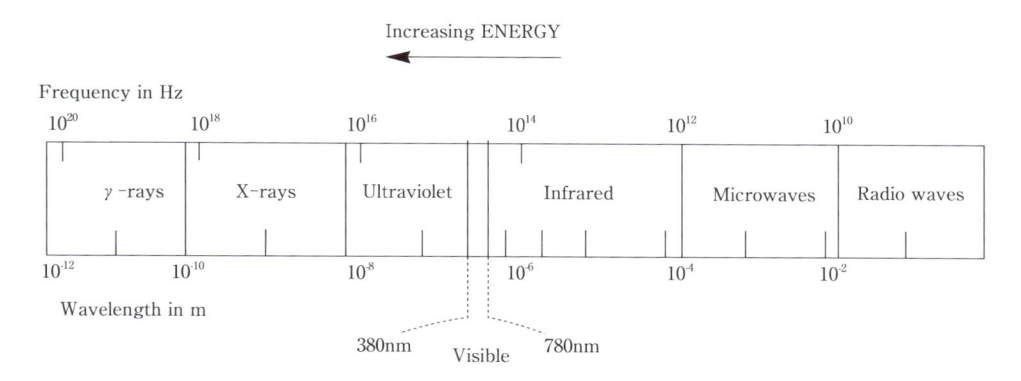

Increasing ENERGY ←

Frequency in Hz: 10^{20}, 10^{18}, 10^{16}, 10^{14}, 10^{12}, 10^{10}

γ-rays | X-rays | Ultraviolet | Infrared | Microwaves | Radio waves

Wavelength in m: 10^{-12}, 10^{-10}, 10^{-8}, 10^{-6}, 10^{-4}, 10^{-2}

380nm Visible 780nm

［注］characterization 特性化（キャラクタリゼーション）, electromagnetic radiation 電磁放射, divide into 〜：〜に分ける, wavelength 波長, not only〜, but — as well：〜だけでなく, —も同様に

2. Electromagnetic radiation is characterized by a frequency, a wavelength, and an amplitude. The energy of light is proportional to its frequency, v, which is the number of waves that pass a fixed point per unit time. More waves mean more energy. Therefore, light of shorter wavelength is higher in energy because more waves can pass a fixed point per unit time. Frequency, which is measured in hertz (Hz), corresponds to a "per second", or s^{-1}, as in "900,000,000 waves per second", or 900 MHz, the operating frequency of most cellular phones.

［注］characterize 特性化する, frequency 周波数, amplitude 振幅, cellular phone 携帯電話

3. Just as matter comes only in discrete units called atoms, electromagnetic "energy" is transmitted only in discrete amounts called quanta. The amount of energy, ε, corresponding to 1 quantum of energy (1 photon) of a given frequency, v, is expressed by the Planck equation

$$\varepsilon = hv$$

where h = Planck's constant (6.62×10^{-34} J·s = 1.58×10^{-34} cal·s).

［注］discrete 不連続の, quantum（単数）量子, quanta（複数）量子

4. The length of a single wave, or wavelength, can be calculated because all waves travel at the same speed. Wave move at the speed of electromagnetic radiation, or as we more commonly say, the speed of light. We abbreviate the speed of light as c, a

constant equal to 3×10^8 m/s. Wavelength, λ, can be calculated from the frequency or energy with the following equations:

$$\lambda = c/\nu, \text{ or } \quad \lambda = hc/\varepsilon$$

注：abbreviate 略する

5. We take advantage of different portions of the electromagnetic spectrum to obtain different types of information about the structure of a molecule.

X-ray Crystallography. High-energy, short wavelength X-rays are passed through a periodic lattice or crystal of molecules. The light diffracts off the planes of atoms in the crystal, and from this scattering, the three-dimensional structures can be calculated.

［注］ take advantage of を利用する，periodic 周期的，lattice 格子，diffract 回折する，scattering 散乱

6. *Ultraviolet (UV) and Visible Spectroscopy.* A broad spectrum (UV-visible range) of light is passed through a sample of molecules. Conjugated networks of π-bonds can absorb this moderate-energy light. Differences in the conjugated π-bond networks lead to the absorption of different region of the spectrum, and from this, the structure can be identified. Often, a single broad and characteristic absorption is diagnostic.

［注］ conjugated 共役した，identify 同定する，diagnostic 診断の

7. *Infrared (IR) Spectroscopy.* This works like UV-visible spectroscopy except that a lower-energy region of the electromagnetic spectrum is used. The portions of the spectrum absorbed by the sample correspond to vibrations of specific bonds in the molecule. Unlike the UV-visible spectrum, the IR spectrum of a molecule contains multiple absorptions that correspond to all the possible vibrations in the molecule. Only a small amount of this information is needed to identify specific functional groups.

［注］ vibration 振動

8. *Nuclear Magnetic Resonance (NMR) Spectroscopy.* When organic molecules containing carbon and hydrogen are placed in a strong magnetic field, radio waves can be used to probe the connectivity of these atoms. From the nuclear magnetic resonance (NMR) spectrum, the carbon and hydrogen arrangements can be determined.

［注］ connectivity 接続性

[Ex. 41][10] Ultraviolet (UV) and Visible Spectroscopy.

1. The ultraviolet (UV) extends from 10^{-8} m to the low-wavelength end of the visible region (4×10^{-7} m). The portion of greatest interest to organic chemists, though, is the narrow range from 2×10^{-7} m to 4×10^{-7} m. Absorption in this region are measured in nanometers (nm), where 1 nm = 10^{-9} m = 10^{-7} cm. Thus, the ultraviolet range of interest is from 200 to 400 nm. The energy of UV radiation corresponds to the amount necessary to raise the energy level of a π-electron in an unsaturated molecule.

［注］ though（副詞）しかし，けれども，necessary to に必要な，unsaturated 不飽和の

2. UV spectra are recorded by irradiating a sample with UV light of continuously changing wavelength. When the wavelength corresponds to the amount of energy required to promote a π-electron in an unsaturated molecule to a higher level, energy is absorbed. The absorption is detected and displayed on a chart that plots wavelength versus the amount of radiation absorbed.

［注］ irradiate 照射する，promote 昇位させる

3. Unlike IR spectra, which generally have many peaks, UV spectra are usually quite simple. Often, there is only a single broad peak, which is identified by noting the wavelength at the very top (λ_{max}). For buta-1,3-diene, λ_{max} = 217 nm.

4. The wavelength of radiation necessary to raise the energy of a π-electron in an unsaturated molecule depends on the nature of the π-electron system in the molecule. One of the most important factors is the extent of conjugation. It turns out that the energy required for an electronic transition decreases as the extent of conjugation increases. Thus, as compared with buta-1,3-diene as given above, hexa-1,3,5-triene absorbs at λ_{max} = 258 nm, and octa-1,3,5,7-tetraene has λ_{max} = 290 nm.

［注］ depend on ～：～に依存する，turn out ～：～が判明する

5. Other kinds of conjugated π-electron systems besides dienes and polyenes also show UV absorptions. Conjugated enones, such as but-3-en-2-one (λ_{max} = 219 nm), and aromatic molecules, such as benzene (λ_{max} = 254 nm), also have characteristic UV absorptions that aid in structure determination.

［注］ besides（前置詞）のほかに，に加えて，cf. beside（前置詞）のそばに，のほかに，

6. UV-visible spectroscopy is a common technique for measuring the concentration of species in solution. Beer's law tells us that the absorbance, A, of a sample is

proportional to the concentration of the molecule in solution.

$$A = \varepsilon c l$$

Here, ε = the molar absorbtivity of the molecule, a molecule-specific constant that reflects how well it absorbs light in units of L mol^{-1} cm^{-1} (usually given for the λ_{max} of the compound), c = concentration of molecules in solution in mol/L, and l = the distance the light travels through the sample, the path-length, in cm.

[注] common 普通の,常用の, concentration 濃度, absorbance 吸光度, molecule-specific 分子特有の, molar absorbtivity モル吸光係数, path-length 光路長

[Ex. 42][10] Infrared (IR) Spectroscopy

1. The IR region covers the range from just above the visible (wavelength 7.8×10^{-7} m) to approximately 10^{-4} m, but only the middle of the region (from 2.5×10^{-4} to 2.5×10^{-3} cm) is used by organic chemists. Instead of wavelength, wavenumber is usually used, and this is given by the reciprocal of the wavelength, in cm^{-1}. Thus, the usual IR spectra extends from 4,000 to 400 cm^{-1}.

[注] wavenumber 波数, reciprocal 逆数

2. Why does a molecule absorb some wavelengths of IR energy but not others? All molecules have a certain amount of energy, which causes bonds to stretch and contract, atoms to wag back and forth, and other molecular motions to occur. The amount of energy a molecule contains is not continuously variable but is quantized. That is, a molecule can vibrate only at specific frequencies corresponding to specific energy levels.

[注] certain ある, stretch 伸ばす, contract 縮める, wag back and forth 前後に振る, quantize 量子化する, specific 特定の, corresponding to~:~に対応する, cause~ to―:~に―させる

3. Take bond stretching, for example. Although we usually speak of bond lengths as if they were fixed, the numbers given are really averages because bonds are constantly stretching and bending, lengthening and contracting. Thus, a typical C―H bond with an average bond length of 110 pm is actually vibrating at a specific frequency, alternately stretching and compressing as if there were a spring connecting the two atoms.

[注] specific 特定の, alternately 交互に, as if~were ―:あたかも~が―かのように, as if there were ―:あたかも―があるかのように

4. When the molecule is irradiated with electromagnetic radiation, energy is absorbed if the frequency of the radiation matches the frequency of the vibration. The result of energy absorption is an increased amplitude for the vibration; in other words, the "spring" connecting the two atoms stretches and compresses a bit further.

Since each frequency absorbed by a molecule corresponds to a specific molecular motion, we can find what kinds of motions a molecule has by measuring the IR spectrum. By then interpreting those motions, we can find out what kinds of bonds (functional groups) are present in the molecule.

　[注]　interpret 解釈する

5. The full interpretation of an IR spectrum is difficult because most organic molecules are so large that they have dozens of different bond stretching and bending motions. Thus, an IR spectrum usually contains dozens of absorptions. Fortunately, we don't need to interpret an IR spectrum fully to get useful information because functional groups have characteristic IR absorptions that don't change from one compound to another: The C = O absorption of a ketone is almost always in the range 1680 to 1750 cm^{-1}, the O — H absorption of an alcohol is almost always in the range 3400 to 3650 cm^{-1}, and so forth. By learning to recognize where characteristic functional group absorptions occur, it is possible to get structural information from IR spectra.

　[注]　interpretation 解釈，stretching 伸縮の，bending 変角の，and so forth など
　　　　さまざまな [to 不定詞，it …to 構文] をチェックしよう

[Ex. 43]$^{10)}$ Magnetic Resonance Imaging (MRI)

1. As practiced by organic chemists, NMR spectroscopy is a powerful method of structure determination. A small amount of sample, typically a few milligrams or less, is dissolved in approximately 1 mL of a suitable solvent, the solution is placed in a thin glass tube, and the tube is placed into the narrow (1-2 cm) gap between the poles of a strong magnet. Imagine, though, that a much larger NMR instrument were available. Instead of a few milligrams, the sample size could be tens of kilograms; instead of a narrow gap between magnet poles, the gap could be large enough for a person to climb into so that an NMR spectrum of body parts could be obtained. What you've just imagined is an instrument for magnetic resonance imaging (MRI), a diagnostic technique of enormous importance in medicine because of its advantage over X-ray or radioactive imaging methods.

　[注]　diagnostic 診断の，advantage 利点

2. Like NMR spectroscopy, MRI takes advantage of the magnetic properties of certain nuclei, typically hydrogen, and of the signals emitted when those nuclei are stimulated by radiofrequency energy. Unlike what happens in NMR spectroscopy, though, MRI instruments use powerful computers and data manipulation techniques to look at the three-dimensional location of magnetic nuclei in the body rather than

at the chemical nature of the nuclei. Most MRI instruments currently look at hydrogen, present in abundance wherever there is water or fat in the body.

　[注]　take advantage of を利用する，certain ある，stimulate 刺激する，manipulation 処理

3. The signals produced vary with the density of hydrogen atoms and with the nature of their surroundings, allowing identification of different types of tissue and even allowing the visualization of motion. For example, the volume of blood leaving the heart in a single stroke can be measured, and heart motion can be observed. Soft tissues that do not show up well on X-rays can be seen clearly, allowing diagnosis of brain tumors, strokes, and other conditions. The technique is also available in diagnosing damage to knees or other joints and is a painless alternative to anthroscopy, in which an endoscope is physically introduced into the knee joint.

　[注]　visualization 可視化，stroke 鼓動，卒中，alternative 代替物，anthroscopy アンスロスコピー（内視鏡を用いる診断），endoscope 内視鏡

4. Several types of atoms in addition to hydrogen can be detected by MRI, and the application of images based on ^{31}P atoms are being explored. The technique holds great promise for studies of metabolism.

　[注]　metabolism 代謝

[Ex. 44][7] Metals

1. The extraction of metals from their ores coincided with the rise of civilization. Bronze, an alloy of copper and tin, was the first metallic material to be widely used. As smelting techniques became more sophisticated, iron became the preferred metal because it is a harder material and more suitable than bronze for swords and plow. For decorative use, gold and silver were easy to use because they are very malleable metals (that is, they can be deformed easily).

　[注]　extraction 採取，抽出，ore 鉱石，alloy 合金，smelt 精錬する，sophisticated 手の込んだ，精巧な，plow 鋤（すき），malleable 展性のある，打ちのばしできる

2. Over the ensuing centuries, the number of metals known climbed to its present large number, the great majority of the elements in the periodic table. Yet in the contemporary world, it is still a small number of metals that dominate our lives, particularly iron, copper, aluminum, and zinc. The metals that we choose must suit the purpose for which we need them, yet the availability of an ore and the cost of extraction are often the main reason why one metal is chosen over another.

　[注]　ensuing 引き続く，contemporary 同時代の，現代の，availability 利用または入手可

能であること

3. High three-dimensional electrical conductivity at SATP (standard ambient temperature and pressure : 25℃ , 100kPa) was the one key characteristic of metallic bonding. Unlike nonmetals, where electron sharing is almost always within discrete molecular units, metal atoms share outer (valence) electrons throughout the metal structure that can be used to explain the high electrical and thermal conductivity of metals together with their high reflectivity.
　［注］　discrete 不連続の，reflectivity 反射性

4. The lack of directional bonding can be used to account for the high malleability and ductility of most metals in that metal atoms can readily slide over one another to form new metallic bonds. The ease of formation of metal bonds account for our ability to sinter the harder metals; that is, we can produce solid metal shapes by filling a mold with metal powder and placing the powder under condition of high temperature and pressure. In those circumstances, metal-metal bonds are formed across the powder grain boundaries without the metal actually bulk melting.
　［注］　malleability 展性，ductility 延性，sinter 焼結する，grain boundary　粒界
　　　　account for 〜: 〜を説明する，in that 〜 : 〜という点で

5. Whereas simple covalent molecules generally have low melting points and ionic compounds have high melting points, metals have melting points ranging from −39℃ for mercury to +3410℃ for tungsten. Metals continue to conduct heat and electricity in their molten state. (In fact, molten alkali metals are often used as heat transfer agents in nuclear power units.) This is evidence that metallic bonding is maintained in the liquid phase.
　［注］　whereas 〜 : 〜の一方で，nuclear power units 原子力設備

6. It is the boiling point that correlates most closely with the strength of the metallic bond. For example, mercury has a boiling point of 357℃ and an enthalpy of atomization of 61 kJ・mol^{-1}, while those of tungsten are 5660℃ and 837 kJ・mol^{-1}, respectively. Thus, the metallic bond in mercury is as weak as some intermolecular forces, whereas that in tungsten is comparable in strength to a multiple covalent bond. In the gas phase, however, metallic elements like lithium exist as pairs, Li$_2$, or, like beryllium, as individual atoms and hence lose their metallic properties. Metals in the gas phase do not even look metallic ; for example, in the gas phase potassium has a green color.
　［注］　―, whereas 〜 : ―，while 〜 : 〜の一方で―，―の一方で〜

be comparable in ― to ～ : ― において ～ と同等である(比較しうるほどである)

7. The simplest metallic bonding model is the electron-sea (or electron-gas) model. In this model, the valence electrons are free to move through the bulk metal structure (hence the term *electron sea*) and even leave metal, thereby producing positive ions. It is valence electrons, then, that convey electric current, and it is the motion of the valence electrons that transfers heat through a metal. However, this model is more qualitative than quantitative. Molecular orbital theory provides a more comprehensive model of metallic bonding. This extension of molecular orbital theory is sometime called band theory.

[注] qualitative 定性的, quantitative 定量的, comprehensive 包括的な, 理解できる
it is～that ― : ― は～(～を強調) : that ― が本(実質)主語節, it(代名詞)は仮(形式)主語

[Ex. 45][7] Antacids

1. One of the major categories of over-the-counter medications is antacids. In fact, the treatment of upset stomachs is a billion-dollar business. Antacids are the most common of the types of inorganic pharmaceuticals. The stomach contains acid ― hydrochloric acid ― since the hydronium ion is an excellent catalyst for the breakdown of complex proteins (hydrolysis) into the simpler peptide units that can be absorbed through the stomach wall. Unfortunately, some people's stomachs overproduce acid. To ameliorate the unpleasant effects of excess acid, a base is required. But the choice of bases is not as simple as in a chemistry lab. For example, ingestion of sodium hydroxide would cause severe and possibly life-threatening throat damage.

[注] over-the-counter medication 店頭市販薬, antacid 制酸剤, upset ひっくりかえった, 具合の悪い, pharmaceuticals 薬品, ameliorate 良くする, ingestion 摂取, throat のど

2. One commonly used remedy for upset stomachs is baking soda, sodium hydrogen carbonate. The hydrogen carbonate ion reacts with hydrogen ion as follows:

$$HCO_3^- + H^+ \longrightarrow H_2O + CO_2$$

The compound has one obvious and one less obvious disadvantage. The compound might increase stomach pH, but it will also lead to the production of gas (so-called flatulence). In addition, extra sodium intake is unwise for those with high blood pressure.

[注] upset ひっくり返った, 具合が悪い, flatulence 腹の張り

3. Some proprietary antacids contain calcium carbonate. This, too, produces carbon dioxide:

$$CaCO_3 + 2H^+ \longrightarrow Ca^{2+} + H_2O + CO_2$$

Although the beneficial aspects of increasing one's calcium intake are mentioned by companies selling such antacid compositions, they rarely mention that calcium ion acts as a constipative. Another popular antacid compound is magnesium hydroxide. This is available in tablet formulation, but it is also marketed as a finely ground solid mixed with colored water to form a slurry called "milk of magnesia". The low solubility of the magnesium hydroxide means that there is a negligible concentration of free hydroxide ion in the suspension. In the stomach, the insoluble base reacts with acid to give a solution of magnesium ion.

$$Mg(OH)_2 + 2H^+ \longrightarrow Mg^{2+} + 2H_2O$$

Whereas calcium ion is a constipative, magnesium ion is a laxative. For this reason, some formulations contain a mixture of calcium carbonate and magnesium hydroxide, balancing the effects of the two ions.

[注] proprietary 専売の, beneficial 有利な, constipative 便秘剤, laxative 便通剤, formulation 処方

[Ex. 46][7) Clathrates, Methane and Carbon Dioxide Hydrates

1. Until a few years ago, clathrates were a laboratory curiosity. Now the methane and carbon dioxide clathrates in particular are becoming of major environmental interest. A clathrate is defined as a substance in which molecules or atoms are trapped within the crystalline framework of other molecules. Here we shall focus on the gas clathrates of water, sometimes called gas hydrates. Though the latter term is widely used, it is not strictly correct, as the term *hydrate* usually implies some intermolecular attraction between the substance and the surrounding water molecules as, for example, in hydrated metal ions.

[注] clathrate クラスレート(包接体), curiosity 好奇心, 珍奇(なもの), framework 骨組, hydrate 水和物, some ある(単数名詞を率いるとき), いくつかの(複数名詞を率いるとき)

2. It was the discovery of methane hydrates on the seafloor that turned clathlates into an issue of major importance. We are now aware that large areas of the ocean floors have thick layers of methane clathrates just beneath the top layer of sediment. It is probable that clathrate layers have formed over eons by the interaction of rising methane from leaking subsurface gas deposits with near-freezing water percolating down through the sediment layers. Each cubic centimeter of hydrate contains about 175 cm^3 of methane gas at SATP (298 K and 100 kPa). The methane content of the

clathrate is sufficient that the "ice" will actually burn. It is believed that the total carbon in the methane clathrate deposits in the world oceans is twice that of the sum of all coal, oil, and natural gas deposits on land.

[注] seafloor 海底，sediment 堆積物，eon = aeon 地質時代の最大区分，永劫，subsurface 表面下の，percolate しみ出る，deposit 堆積床，鉱床，it was ～ that ―：―したのは，～であった（強調），it～that ―：―は～：that ― が本主語節，it(代名詞)は仮主語，twice that それ（先行する the total carbon の代名詞）の2倍，interaction of～with ―：～と―との相互作用，sufficient that～：～するに十分な

3. Because the stability of methane clathrates is so temperature and pressure dependent, there is concern that the warming of oceans may lead to the melting of chlathrate deposits, releasing large volumes of methane into the atmosphere. The released methane would then have a significant effect on climate because methane is a potent greenhouse gas. It has been argued that some sudden past changes of climate were triggered by methane release from clathrates. For example, the lowering of water levels during ice ages would have reduced the pressure on seabed deposits, possibly liberating large volume of gas. The increased methane levels would then have caused global warming, terminating the ice age.

[注] ～ dependent～依存性の，potent 有力な，greenhouse 温室，trigger 触発する，引金を引く，that ～（同格節を率いる接続詞）～という，
it ～ that ―：― は ～：that ― が本主語節，it(代名詞)は仮主語
some ある（単数名詞を率いるとき），いくつかの（複数名詞を率いるとき）

4. Deep ocean sequestration of carbon dioxide has been suggested as one possible method of storing waste carbon dioxide produced by power plants and industrial processes. When carbon dioxide is released into the deep ocean, under the ambient conditions of temperature and pressure, it forms a solid clathrate. The clathrate has a high stability; for example, at a depth of 250 m, where the pressure is about 2.7 MPa, the clathrate is stable at +5℃. Whereas "normal" ice is less dense than liquid water, the carbon dioxide clathrate has a density of about $1.1 \text{ g} \cdot \text{cm}^{-3}$ and sinks to the ocean floor. It has been proposed that megatons of excess carbon dioxide could be disposed of in this way.

[注] sequestration 隔離，megatons 数メガトン（10^6 ton）

5. There are three major concerns with this concept. First and most important, the layer of carbon dioxide clathrate will smother the exotic bottom life of the deep oceans where the clathrates are deposited. Second, experiments have already shown that fishes exhibit respiratory distress when they approach the carbon dioxide-saturated water around experimental clathrate deposits. Third, over an extended

period, perhaps hundreds of thousands of years, the clathrates will probably release their captive carbon dioxide to the surrounding waters, causing a pH decrease of the oceans. The pH change would obviously have an effect on the ecological balance of marine life.

　［注］　smother 窒息させる，respiratory distress 呼吸困難，X-saturated : X を飽和した，captive 捕獲（包接）した，that 〜 : 〜 ということを（目的語節を率いる接続詞）

[Ex. 47][7] Silicon, Silica, and Zeolites

1. About 27 percent by mass of the Earth's crust is silicon. However, silicon itself is never found in nature as the free element but only in compounds containing oxygen-silicon bonds. The element is a gray, metallic-looking, crystalline solid. Although it looks metallic, it is not classified as a metal because it has a low electrical conductivity.

　［注］　crust 殻，electrical conductivity 電気伝導度

2. About half a million tons per year of silicon are used in the preparation of metal alloys. Although alloy manufacture is the major use, silicon plays a crucial role in our lives as the semiconductor that enables computers to function. The purity level of the silicon used in the electronics industry has to be exceedingly high. For example, the presence of only 1 ppb of phosphorus is enough to drop the specific resistance of silicon from 150 to 0.1 kΩ・cm. As a result of the expensive purification process, ultrapure electronic grade silicon sells for over 1000 times the price of metallurgical grade (98% pure) silicon.

　［注］　crucial 決定的な，semiconductor 半導体，exceedingly 非常に，ppb = part per billion 10 億分の 1（$1/10^9$），specific resistance 比抵抗

3. The most common crystalline form of silicon dioxide, SiO_2, commonly called silica, is the mineral quartz. Most sands consist of particles of silica that usually contain impurities such as iron oxides. It is interesting to note that carbon dioxide and silicon dioxide share the same type of formula, yet their properties are very different. Carbon dioxide is a colorless gas at room temperature, whereas solid silicon dioxide melts at 1600℃ and boils at 2230℃. The difference is due to bonding factors. Carbon dioxide consists of small, triatomic, nonpolar molecular units whose attraction to one another is due to dispersion forces. By contrast, silicon dioxide contains a network of silicon-oxygen covalent bonds in a giant molecular lattice. Each silicon atom is bonded to four oxygen atoms, and each oxygen atom is bonded to two silicon atoms, an arrangement consistent with the SiO_2 stoichiometry of the compound.

　［注］　triatomic 3 原子の，one another お互いに，dispersion force 分散力，stoichiometry

化学量論

4. At first thought, one might consider that aluminum, a metal, and a silicon, a semimetal/nonmetal, have little in common. However, in a large number of mineral structures aluminum partially replaces silicon. This should not be too surprising, for aluminum and silicon fit in similar-sized cation sites. Of course, this is premising the bonding is ionic.

　[注]　in common 共通に，replace 置き換える，premise 前提とする

5. The large range of aluminosilicates is actually derived from the basic silica structure of a three-dimensional array of SiO_4 units linked by the corner oxygen atoms. In silica, the structure will be neutral. Then Al^{3+} is substituted for Si^{4+}, the lattice acquires one net negative charge for every replacement. For example, replacement of one-fourth of the silicon atoms by aluminum results in an anion of empirical formula $[AlSi_3O_8]^-$: replacement of one-half the silicon atoms gives the formula $[Al_2Si_2O_8]^{2-}$. The charge is counterbalanced by Group 1 or 2 cations. This particular family of minerals comprises the feldspars, components of granite. Typical examples are orthoclase $K[AlSi_3O_8]$ and anorthite $Ca[Al_2Si_2O_8]$.

　[注]　aluminosilicate アルミノ珪酸塩，replacement 置換，result in の結果となる，counterbalance つり合わせる，相殺する，comprise 構成する，feldspar 長石，granite 花崗岩，orthoclase 正長石，anorthite 灰長石

6. One three-dimensional aluminosilicate structure has open channels throughout the network. Compounds with this structure are known as zeolites, and their industrial importance is skyrocketing. A number of zeolites exist in nature, but chemists have mounted a massive search for zeolites with novel cavities throughout their structure. There are four major uses for zeolites : (1) as ion exchangers, (2) as adsorption agents, (3) for gas absorption, and (4) as industrial catalysts.

　[注]　channel 溝，水路，skyrocket 急騰する，mount 登る，始める，cavity 穴

7. One of the most important catalysts is $Na_3[(AlO_2)_3(SiO_2)] \cdot xH_2O$, commonly called ZSM-5. This compound does not occur in nature ; it was first synthesized by research chemists at Mobil Oil. It is higher in aluminum than most naturally occurring zeolites, and its ability to function depends on the high acidity of water molecules bound to the high-charge-density aluminum ions. In fact, the hydrogen in ZSM-5 is as strong a Brønstead-Lowry acid as that in sulfuric acid. ZSM-5 catalyzes reactions by admitting molecules of the appropriate size and shape into its pores and

then acting as a strong acid. An example is synthesis of ethylbenzene, an important organic reagent, from benzene and ethylene. It is believed that ethylene is first protonated within the zeolite to yield ethyl cation, which attacks benzene to give the product.

[注] admit 容れる，as strong as と同じぐらい強い，that 代名詞（hydrogen の代わり）

[Ex. 48]15) Ions in Your Body

1. Of the 90 elements naturally found on earth, 25 are essential to living organisms. The major elements in all living organisms are hydrogen, oxygen, carbon, and nitrogen — they are found in organic compounds such as carbohydrates, fats, proteins, and vitamins. The remaining 21 elements are minerals — these are naturally occurring inorganic elements. The minerals can be divided into two groups, the major minerals and the trace minerals.

[注] of の内，essential 必須の

2. The major minerals, or macrominerals, are found in greater mass and include calcium, phosphorus, potassium, sulfur, chloride, sodium, and magnesium. Together, calcium and phosphorus make up three-fourths of the mass of all the minerals present in your body. The trace minerals are iron, iodide, fluoride, manganese, zinc, molybdenium, copper, cobalt, chromium, selenium, arsenic, nickel, silicon, and boron. Even though the trace minerals are found in lower amounts than the major minerals, they are equally important to your body. A daily deficiency of a few micrograms of iodine is just as serious as a defficiency of several hundred milligrams of calcium.

[注] make up を成す，deficiency 不足

3. The element iodine usually found in nature as the iodide ion (I⁻). Iodide ion is needed by the body in an extremely small quantity, but obtaining this quantity is critical. The body uses iodide ion to make thyroxin, the hormone responsible for regulating the basal metabolic rate. Without iodide ion, the body cannot make thyroxin. Thyroxin is synthesized on the thyroid, a gland located in the lower neck. When the iodide level of the blood is low, the cells of the thyroid enlarge, forming a goiter. People with this condition suffer slugginess and weight gain. Infants born to mothers suffering this condition may be born with irreversible mental and physical retardation known as cretinism.

[注] thyroxin チロキシン，basal metabolic rate 基礎代謝率，thyroid 甲状腺，gland リンパ腺，goiter 甲状腺腫，slugginess 無精，cretinism クレチン病

4. Calcium is essential to bone formation, tooth formation, nerve transmission, maintenance of normal blood pressure, blood clotting, muscle contraction, and heart

function. Cells need continuous access to calcium, so the body maintains a calcium ion concentration in the blood. The skeleton serves as a bank from which the blood can borrow and return calcium. You can go without adequate calcium for years without suffering noticeable symptons and then, late in life discover that your calcium savings are depleted and the integrity of your skeleton can no longer be maintained. Adult bone loss, or osteoporosis, is a health problem for many older people. Each year over a million people in the U. S. suffer bone breaks due to osteoporosis. Menopause increases bone loss in women.

[注] clotting 凝固，deplete 枯渇させる，integrity 元のままの状態，osteoporosis 骨粗しょう症，menopause 閉経

5. The element phosphorus is needed primarily as phosphate ion, PO_4^{3-}. Together with calcium ion, phosphate ion is essential to bone and tooth formation. Phosphates are important constituents of buffers in the blood which maintain blood acid-base balance. Phosphorus compounds are essential to energy transfer in cells via adenosine triphosphate (ATP). Phosphorus is an essential part of the genetic materials DNA and RNA. Certain lipids — phospholipids — that form the membranes around each cell also contain phosphorus.

[注] phospholipid リン脂質

6. Iron is primarily needed as a component of hemoglobin and myoglobin. Both compounds use iron to carry and hold oxygen. Hemoglobin is the oxygen carrier in the blood and myoglobin is the oxygen reservoir in the muscle cells. Iron is also used by many enzymes in the metabolic transfer of energy.

7. Buffers are solutions that resist a change in pH. They protect against drastic changes in pH when acids or bases are added to the solution. The body's blood and extracellular fluids contain buffers. The buffering capacity of a system may be overloaded if huge amounts of acid or base are added; in the blood, this can lead to acidosis or alkalosis.

[注] acidosis アシドーシス（酸性血症），alkalosis アルカローシス（アルカリ性血症）

8. A mixture of potassium dihydrogenphosphate (KH_2PO_4) and potassium monohydrogenphosphate (K_2HPO_4) involves the equilibrium as follows：

$$H_2PO_4^- + H_2O = HPO_4^{2-} + H_3O^+$$

If acid is added to this buffer solution, the stress of added hydronium ion causes the equilibrium to shift to the left, relieving the stress. Most of the additional hydronium ion is changed into water and the pH stays relatively constant. Bases decrease the concentration of hydronium ion in a solution. If base is added to the buffer solution, the stress of reduced hydronium ion concentration causes the equilibrium to shift to the right. Water is converted to hydronium ion and the pH does not significantly change.

[Ex. 49][10] Acid and Base

1. Acidity and basicity are related to electronegativity and bond polarity. Acid-base behavior of organic molecules helps explain much of their chemistry. Two definitions of acidity are frequently used: Brønstead-Lowry definition and Lewis definition.

[注] electronegativity 電気陰性度，bond polarity 結合極性，definition 定義，
be related to に関係がある，help explain = help to explain

2. A Brønstead-Lowry acid is a substance that donates a hydrogen ion (or proton H^+), and a Brønstead-Lowry base is a substance that accepts a hydrogen ion (or proton H^+). Acids react with bases: when gaseous hydrochloric acid dissolves in water, the acid, HCl, donates a proton to the base, a water molecule. The products of this reaction are a new acid and base. To distinguish the products from the reactants, we identify the products as a conjugate acid and a conjugate base. The hydronium ion (H_3O^+) produced is the conjugate acid because it can donate a proton, and the chloride ion (Cl^-) is the conjugate base because it can accept a proton.

$$H-Cl + H_2O \longrightarrow Cl^- + H_3O^+$$
$$\text{Acid} \quad \text{Base} \quad \text{Conjugate base} \quad \text{Conjugate acid}$$

[注] donate 与える，accept 受ける，dissolve 溶ける，product 生成物，conjugate 共役の

3. The ability of an acid to donate a proton depends on the acid. Stronger acids such as HCl react almost completely with water, whereas weaker acids such as acetic acid (CH_3COOH) react only slightly. The exact strength of a given acid in water solution can be expressed by its acidity constant, K_a. For the reaction of any generalized acid HA with water, the equilibrium and the acidity constant, K_a, are written as follows:

$$HA + H_2O \rightleftarrows A^- + H_3O^+$$

$$K_a = \frac{[H_3O^+][A^-]}{[HA]}$$

where the brackets [ ] refer to the concentration of the closed species in molarity (M), or moles per liter (mol/L). Stronger acids have their equilibria toward the right and thus have larger acidity constants; weaker acids have their equilibria toward the left and have smaller acidity constant.

［注］　whereas 一方，acidity constant 酸性度定数，refer to を指す，を言う，concentration 濃度，closed 括弧内に入れた，equilibrium（単数），equilibria（複数）平衡

4. The range of K_a values for different acids is enormous, running from about 10^{15} for the strongest acids to about 10^{-60} for the weakest. The common inorganic acids such as H_2SO_4, HNO_3, and HCl have K_a's in the range 10^2 to 10^9, while many organic acids have K_a's in the range of 10^{-5} to 10^{-15}. As you gain more experience, you'll develop a rough feeling for which acids are "strong" and which are "weak" (remembering that the terms are always relative).

［注］　while 一方

5. Acid strengths are normally given using pK_a values, where the pK_a is equal to the negative common logarithm of the K_a,

$$pK_a = -\log K_a$$

A stronger acid (larger K_a) has a smaller pK_a, and a weaker acid (smaller K_a) has a larger pK_a. Some common acids in increasing order of strength have the pK_a values as follows : water (H_2O) 15.74, acetic acid (CH_3CO_2H) 4.76, hydrofluoric acid (HF) 3.45, hydrochloric acid (HCl) -7.0

［注］　common logarithm 常用対数，common 普通の

6. The Lewis definition of acids and bases differs from the Brønstead-Lowry definition in that it's not limited to substances that donate or accept protons. A Lewis acid is a substance that has a vacant valence orbital and can thus accept an electron pair; a Lewis base is a substance that donates an electron pair. The donated pair of electrons is shared between Lewis acid and base in a newly formed covalent bond.

$$A \ + \ :B \longrightarrow A—B$$
　　　　　　　Lewis acid　　Lewis base

［注］　valence orbital 原子価軌道，covalent bond 共有結合

7. A proton (H^+) is a Lewis acid because it accepts a pair of electrons to fill its vacant 1s orbital when it bonds to a base. However, Lewis acids include not only proton donors but many other species as well. A compound such as aluminum trichloride ($AlCl_3$) is a Lewis acid because it too accepts an electron pair from a Lewis base such as trimethylamine ($N(CH_3)_3$) to fill a vacant valence orbital.

[注] not only～but(also)—(as well)：～だけでなく，—も

[Ex. 50][7] Superacids

1. A superacid can be defined as an acid that is stronger than 100 % sulfuric acid. In fact, chemists have synthesized superacids that are from 10^7 to 10^{19} times stronger than sulfuric acid. A common Brønstead superacid is perchloric acid. When perchloric acid is mixed with pure sulfuric acid, the sulfuric acid actually acts like a base:

$$HClO_4 + H_2SO_4 \rightleftharpoons H_3SO_4^+ + ClO_4^-$$

Fluorosulfuric acid, FSO_3H, is the strongest Brønstead superacid; it is more than 1000 times more acidic than sulfuric acid. This superacid is an ideal solvent because it is liquid from $-89℃$ to $+164℃$.

[注] superacid 超強酸(スーパーアシッド)，perchloric acid 過塩素酸

2. A Brønstead-Lewis superacid is a mixture of a powerful Lewis acid and a strong Brønstead superacid. The most potent combination is a 10% solution of antimony pentafluoride, SbF_5, in fluorosulfuric acid. The addition of SbF_5 increases the acidity of FSO_3H several thousand times. The reaction between the two acids is very complex, but the super-hydrogen-ion donor in the mixture is the $FSO_3H_2^+$ ion. This acid mixture will react with many substances, such as hydrocarbons, that do not react with normal acids. For example, propene, C_3H_6, reacts with this ion to give the propyl cation:

$$C_3H_6 + FSO_3H_2^+ \longrightarrow C_3H_7^+ + FSO_3H$$

3. The solution of SbF_5 in FSO_3H is commonly called "Magic Acid". The name originated in the Case Western Reserve University laboratory of George Olah, a pioneer in the field of superacid (and recipient of the Nobel Prize in chemistry in 1994). A researcher working with Olah put a small piece of Christmas candle left over from a lab party into the acid and found that it dissolved rapidly. He studied the

resulting solution and found that the long-chain hydrocarbon molecules of the paraffin wax had added hydrogen ions and the resulting cations had rearranged themselves to form branched-chain molecules. This unexpected finding suggested the name "Magic Acid", and it is now a registered trade name for the compound. This family of superacids is used in the petroleum industry for the conversion of the less important straight-chain hydrocarbons to the more valuable branched-chain molecules, which are needed to produce high-octane gasoline.

［注］　rearrange 転移する

[Ex.51][10] Nature of Organic Molecules

1. There are more than 27 million organic molecules, each of which has its own unique physical and chemical properties. Some of these compounds occur in nature. Others are the creation of chemists. The "rules" that nature and people use to make organic molecules rest largely on understanding the chemistry of small combinations of atoms. So, instead of 27 million compounds with random reactivity, there are a few dozen families of compounds whose chemistry is reasonably predictable. We'll study the chemistry of the most common families.

［注］　creation 創造，rest on に頼る，nature 自然，天然，性質

2. The structural features that make it possible to classify compounds by reactivity are called functional groups. A functional group is a group of atoms within a larger molecule that has a characteristic chemical behavior. Chemically, a given functional group behaves almost the same way in every molecule it's in. For example, one of the simplest functional groups is the carbon — carbon double bond. Because the electronic structure of the carbon — carbon double bond remains essentially the same in all molecules where it occurs, its chemical reactivity also remains the same. For instance, ethylene, the simplest compound with a carbon — carbon double bond, undergoes reactions that are identical to those of menthene, a substantially larger molecule found in peppermint oil. Both, for example, react with Br_2 to give products in which a bromine atom has added to each of the double-bond carbons. The chemistry of every organic molecule, regardless of size and complexity, is determined by the functional group it contains.

注：structural feature 構造的特長，functional group 官能基，identical 同じ，
　　it(代名詞) = to classify compounds…

[Ex. 52] [15)] Aspirin

1. Ancient native Americans used the bark of the willow tree to counter fever and pain. Europians learned of the medicinal properties of willow bark in 1763 when clergyman Edward Stone read a paper to the Royal Sosiety of London. Willow bark extract was eventually found to be a powerful analgesic (pain reliever), antipyretic (fever reducer), and anti-inflammatory (reduces swelling) drug. Organic chemists were able to isolate and identify the active ingredient in willow bark in 1838. The ingredient, salicylic acid, was named from salix, Latin for the willow tree. The use of salicylic acid was limited because its acidic properties caused severe irritation of the stomack.

［注］ bark 皮，willow 柳，clergyman 牧師，extract 抽出物，analgesic 鎮痛剤，antipyretic 解熱剤，anti-inflammatory 炎症抑制剤，swelling 腫れ，salix やなぎ属の総称，identify 同定する，ingredient 成分，irritation 刺激，be able to ─：─できる

2. The German chemist Felix Hoffman, working for the firm of Bayer, synthesized an ester of salicylic acid, acetylsalicylic acid in 1893. Acetylsalicylic acid was marketed by Bayer under the trade name of aspirin. During World War I the U. S. Government seized Bayer's assets and sold them and the Bayer name to Sterling Products (now Sterling Winthrop Inc.).

［注］ seize 捕らえる，差し押さえる，asset 資産

3. Like salicylic acid, aspirin is analgesic, antipyretic, and anti-inflammatory; it is much less irritating to the stomach than salicylic acid. It relieves headache, toothache, pain and fever of colds, muscle aches, menstrual pain, and pain of arthritis. Studies have suggested that one low dose aspirin tablet per day (contains 81 mg of acetylsalicylic acid) may reduce the risk of heart attack, stroke, and colon cancer.

［注］ arthritis 関節炎，colon 結腸

4. Aspirin does cause slight gastrointestinal bleeding that can, over time, cause iron deficiency or gastric ulcers. These complications can be avoided with enteric-coated aspirin, which does not dissolve until reaching the small intestine. Aspirin should not be given to children who have influenza or chicken pox because of the risk of the rare and often fatal Reye's syndrome ; children suffeing from these deseases should be given pain reliever such as acetaminophen. Unless directed by a physician, aspirin should not be taken during the last 3 months of pregnancy.

［注］ gastrointestinal bleeding 胃腸の出血，small intestine 小腸，gastric ulcer 胃潰瘍，complication 併発症，enteric-coated 腸溶性コートされた，chicken pox 水ぼうそう，Reye's syndrome ライ症候群，pregnancy 妊娠

5. Only in the last few years biochemists have begun to understand how aspirin works its wonders. Aspirin prevents the formation of prostaglandins, a class of compounds responsible for evoking pain, fever, and local inflammation. Like many medicinal drugs, aspirin was developed from a naturally occurring substance. Chemists first isolated the active ingredient, detemined its structure, and then improved on the original. Further improvement is possible once the exact mechanism of aspirin's interaction with prostaglandins is understood.

［注］ responsible for の原因となる，の責任がある

[Ex. 53][10] Sweeteners

1. Say the word 'sugar' and most people immediately think of sweet-tasting candies, deserts, and such. In fact, most simple carbohydrates do taste sweet, but the degree of sweetness varies greatly from one sugar to another. With sucrose (table sugar) as a reference point, fructose is nearly twice as sweet, but lactose is only about one-sixth as sweet. Comparisons are difficult, though, because sweetness is a matter of taste and the ranking of sugars is a matter of personal opinion.

［注］ immediately 直ちに，and such その他同様なもの（人）

2. The desire of many people to cut their caloric intake has led to the development of synthetic sweeteners such as saccharin, aspartame, and acesulfame. All are far

sweeter, more than about 180 times sweeter, than natural sugars, so the choice of one or another depends on personal taste, government regulations, and (for baked goods) heat stability. Saccharin, the oldest synthetic sweetener, has been used for more than a century, although it has a somewhat metallic aftertaste. Doubts about its safety and potential carcinogenicity were raised in the early 1970s, but it has now been cleared of suspicion. Acesulfame potassium, the most recently approved sweetener, is proving to be extremely popular in soft drinks because it has little aftertaste. None of the three synthetic sweeteners has any structural resemblance to a carbohydrate.

Saccharin　　Aspartame　　Acesulfame-K

[注]　acesulfame-K アセスルファム・カリウム，depend on〜：〜に依存する，aftertaste あと味，carcinogenicity 発がん性，suspicion 疑い

3.[20]　Saccharin, the first artificial sweetener, was discovered over a hundred years ago, long before it became fashionable to use a substitute for the common table sugar. This happened in the laboratory of Ira Remsen, the most famous American chemist of the 19th century.

　Remsen was born in New York in 1846 ; he went to Germany for graduate study at the universities of Munich, Göttingen, and Tübingen. Returning to the United States, he became professor of chemistry at Williams College, then at Johns Hopkins University. He established the first chemistry department in U. S. of a quality equal to those of Europe and he counted among his students many future leading American chemists. He later became president of John Hopkins University. One of Remsen's students was my scientific "great-great-grandfather" : E. P. Kohler. One of his students was James B. Conant, whose student was Louis F. Fieser, whose student was Charles C. Price, whose student was Royston M. Roberts. Roberts (the author of this article) sometimes liked to point out that he can trace his chemical ancestry back to Wöhler, the father of organic chemistry because Remsen was a student of Rudolph Fittig, who was a student of Friedrich Wöhler.

[注]　合成甘味料に関して，3〜5は，その発見のセレンディピティの逸話を，R. M. Roberts の本[20]から引用した。
artificial 人工の，substitute 代替品，establish 設立する，great-great-grandfather 曽曽祖父（祖父よりも2代前の），ancestry 祖先

4.[20] In 1879, one of Remsen's associates was pursuing a problem assigned as part of an ongoing theoretical research program. While doing this, the associate, who was named Fahlberg, noticed that a substance that he had prepared and accidentally spilled onto his hand tasted unusually sweet. (Chemists were not nearly so cautious then about smelling and tasting the materials they worked with as they are now.) Fahlberg apparently foresaw the possible importance of the new sweet-tasting substance, because he developed a commercial synthesis and took out a patent on it in 1885. The name he chose for it was saccharin, from the Latin word for sugar, saccharum.

[注] associate 同僚，仲間，ongoing 進行中の，apparently～：見かけ上～，～と思われる，foresee 予見する

5.[20] Aspartame is chemically a methyl ester of a dipeptide, L-aspartyl-L-phenylalanine, which was also found to be a strong sweetener entirely by accident by a chemist James M. Schlatter. The sweet taste of aspartame could not have been predicted from a knowledge of the properties of the component amino acids — one of them has a "flat" taste and the other is bitter. The extremely sweet taste that resulted from the combination of the two and conversion to the methyl ester was a complete surprise.

Schlatter describes in his book (1984) about the actual discovery as follows. "In December 1965, I was heating aspartame in a flask with methanol when the mixture bumped onto the outside of the flask. As a result, some of the powder got onto my fingers. At a slightly later stage, when licking my finger to pick up a piece of paper, I noticed a very strong, sweet taste…" Unlike saccharin, which is excreted unchanged, aspartame is metabolized into its constituent natural amino acids, which are further metabolized by the usual body pathways. Because Schlatter knew this much about the metabolism of peptides, he was bold enough to taste the material that had splashed onto the outside of his flask.

[注] excrete 排泄する，metabolize 代謝する，splash はね飛ぶ，one, the other 一方，他方，enough to に十分な

[Ex. 54][10] Terpenes

1. It has been known for centuries that distillation of many plant materials with steam produces a fragrant mixture of liquids called essential oils. For hundreds of years, such plant extracts have been used as medicines, spices, and perfumes. The investigation of essential oils also played a major role in emergence of organic chemistry as a science during the 19th century.

[注] essential oil 精油，extract 抽出物，emergence 出現

2. Chemically, plant essential oils consist largely of mixtures of compounds called terpenes — small organic molecules with an immense diversity of structure. Thousands of different terpenes are known, and many have carbon — carbon double bonds. Some are hydrocarbons, and others contain oxygen; some are open-chain molecules, and others contain rings. For example,

Myrcene (oil of bay)　　α-Pinene (turpentine)　　Carvone (spearmint oil)

［注］ consist of から成る，diversity 多様性，open-chain 開鎖の

3. All terpenes are related, regardless of their apparent structural differences. According to a formalism called the isoprene rule, terpenes can be thought of as arising from head-to-tail joining of five-carbon isoprene (2-methylbuta-1, 3-diene) units. Carbon 1 is the head of the isoprene unit, and carbon 4 is the tail. For example, myrcene contains two isoprene units joined head to tail, forming an eight-carbon chain with 2 one-carbon branches. α-Pinene similarly contains two isoprene units assembled into a more complex cyclic structure.

Isoprene　　　　Myrcene

［注］ regardless of に関わらず，apparent 見かけの，according to によれば，head-to-tail 頭―尾

4. Terpenes are classified according to the number of isoprene units they contain. Thus, monoterpenes are 10-carbon substances from two isoprene units, sesquiterpenes are 15-carbon molecules from three isoprene units, diterpenes are 20-carbon substances from four isoprene units, and so on. Monoterpenes and sesquiterpenes are found primarily in plants, but the higher terpenes occur in both plants and animals and many have important biological roles. The triterpene lanosterol, for example, is the precursor from which steroid hormones are made.

Lanosterol, a triterpene (C_{30})

［注］ precursor 前駆体, sesquiterpene セスキテルピン （sesqui- は 1.5 の意味の接頭語；
cf. ethylaluminium sesquichloride : $Et_{1.5}AlCl_{1.5} = Et_3Al_2Cl_3$）

5. Research has shown that isoprene itself is not the true biological precursor of terpenes. Nature instead uses two "isoprene equivalents" — isopentenyl diphosphate and dimethylallyl diphosphate — five-carbon molecules that are themselves made from acetic acid. Every step in the biological conversion from acetic acid through lanosterol to human steroids has been worked out — an immense achievement for which several Novel Prizes have been awarded.

$3\ H_3C-\overset{O}{\underset{\|}{C}}-OH \Rightarrow$ Isopentenyl diphosphate / Dimethylallyl diphosphate

［注］ immense 膨大な, achievement 偉業, 達成

[Ex. 55] [10] Drugs

1. Where do drugs come from? The answer to this question might be "land", "sea", "hard work", or "luck", depending on whom we ask. Historically, the origins of drugs are medicinal plants. Hippocrates recommended chewing the bark of a willow tree to relieve toothaches. More than 2000 years later, a German chemist at Bayer made, in the laboratory, aspirin, a derivative of the same active ingredient in the bark of willow tree — salicylic acid.

［注］ depending on によって, 依存して

2. Indeed, whether it is a drug for malaria or the most exciting new cancer therapy, nature often provides the lead. In fact 67% of the 1031 new drug leads reported to the U. S. Food and Drug Administration (FDA) between 1981 and 2002 were derived from nature, according to a study conducted by the U. S. National Cancer Institute. (However, as we'll see shortly, only a small fraction of these molecules become medicines.)

[注] according to によれば

3. The origins of these drugs can be depicted in the four parts as follows. First (15%), as a relatively new and growing source of drugs, the vaccines and protein and peptide drugs are produced in fermentation using genetically engineered bacteria and other organisms. Second part (28%) is the molecules used exactly as they are found in nature. For economic reasons, these molecules might be produced either by nature and subsequently isolated or prepared synthetically in the laboratory. Third part (24%) is the molecules that do not occur in nature, but the active part of the molecules is taken from nature and incorporated into an unnatural molecule. Fourth (33%) comes from the molecules that do not have natural origins and are prepared by chemists in the laboratory.

[注] fermentation 発酵，genetically engineered 遺伝子工学の，either A or B：AかあるいはBか

4. With all these sources of compounds, why are medicines so slow to come to market and so costly when they arrive? The answer is time. Once scientists identify a potential drug, it takes, on average, 11 years for the drug to come to market. These 11 years can be divided into different phases. Preclinical development focuses primarily on scientific and economic issues. The drug must be safe in animals and cheap enough to produce at the large scales required for human (clinical) trials.

[注] preclinical 前臨床の

5. In the first phase of clinical trials, the drug is administrated to healthy people, who are then monitored for side effects. In the second phase the drug is administrated to a relatively small number of patients who have the disease or condition the drug is expected to improve or cure. The third phase studies allow a broader range of patients to take the drug. If these studies are successful, the company will request approval for widespread clinical use from the FDA. A newly approved drug can be worth billions of dollars to a company.

[注] administrate 投与する，side effect 副作用

6. The third phase still does not ensure a successful product. At times, small populations of patients develop unfavorable and serious side effects after the drug has been in widespread circulation for a few years, leading the company to withdraw the drug from the market. In addition, drugs do not always produce a significant profit. One reason is that competitors constantly introduce new drugs. Another reason is that the patents protecting these drugs are good for only 20 years. Since

patents are filed during the preclinical trials, the company has exclusive rights to the drug for only about 6 years. Afterward, generic drug manufacturers can enter the market and sell the drug for significantly less money because these companies have no expenses associated with preclinical or clinical trials.

[注] widespread circulation 広汎な普及, competitor 競合者, exclusive right 占有権, generic drug manufacturer ジェネリック薬品製造会社, significantly 意味あるぐらいに, かなり, not always いつも…ではない, that ということ (competers 以降の補語節を率いる接続詞)

[Ex. 56]¹⁵⁾ Alkaloids

1. Alkaloids are naturally occurring nitrogen compounds having pronounced physiological activity. Like other simpler amines, alkaloids are bases; the name alkaloid comes from their alkaline properties. These complex molecules come from plants and many have medical uses. They generally are insoluble in water, and are often found in commercial products in the form of their water-soluble acid salts.

$$\text{alkaloid (water insoluble)} + \text{acid} \longrightarrow \text{alkaloid acid salt (water soluble)}$$

The pure alkaloid can be obtained as a precipitate from its acid salt by treatment with a base such as sodium hydroxide.

$$\text{alkaloid acid salt (water solution)} + \text{base} \longrightarrow \text{alkaloid (water insoluble)} \downarrow$$

[注] pronounced 著しい, physiological 生理学の, precipitate 沈殿

2. Among the better known alkaloids are caffeine and quinine. Caffeine is found in coffee, tea, and kola nuts. Caffeine is a mild stimulant to the central nervous system causing increased alertness and the ability to put off sleep. Coffee contains 2-5% caffeine, about the same amount present in tea leaves. Many people prefer to drink decaffeinated coffee. Caffeine is extracted from the beans prior to roasting alkaloid by use of carbon dioxide liquid as a safe solvent

[注] stimulant 刺激剤, central nervous system 中枢神経系, alertness 警戒心, prior to の前に

3. Cola soft drink beverages are made from the extract of the kola (cola) nut, which contains caffeine. Manufacturers add phosphoric acid, caramel, sweeteners, and carbonated water. They usually remove all the caffeine from the kola extract and then add the correct amount as required by the U. S. Food and Drug Administration (FDA). Caffeine is also added to some energy drinks, which are not included in FDA regulations, so they may have caffeine limits above those of the cola and soda drinks.

[注] extract 抽出物

4. Caffeine has little effect on the blood pressure in minute quantities but large amounts increase blood pressure. Individuals who drink coffee, tea, or cola drinks in large amounts can develop both a tolerance for and a dependence on caffeine. Heavy users can experience withdrawal symptoms of lethargy, headache, and even nausea after 18 hours of abstinence.

　［注］　minute 微小の，tolerance 抗薬力，寛容，withdrawal symtom 禁断症状，lethargy 無気力，nausea 吐き気，abstinence 禁欲

5. The structure of caffeine is very similar to that of adenine and guanine — important components of the genetic material DNA. This similarity has raised concern that caffeine might cause cancer or birth defects. So far there is little evidence to support this concern. Some people consider caffeine to be an addictive drug and some religions prohibit use of beverage containing caffeine.

　［注］　addictive 習慣性の

6. Quinine is an antipyretic (fever-reducer) and was for a long time the only known remedy for malaria. Quinine binds to the DNA of malaria-infected cells and inhibits their replication. Only infected cells are affected because they absorb quinine in higher concentration than unaffected cells. The alkaloid is found in the bark of the cinchona tree ; these trees were extensively cultivated in Indonesia in the late 19th century.

　［注］　infect 感染させる，replication 複製，cinchona キナの木

7. When the Japanese invasion of Indonesia in World War II cut off the supply of quinine needed by Allied troops, American Chemist Robert Woodward succeeded in synthesizing quinine from coal tar. Woodward was renown for his ability to synthesize complex organic substances and was awarded the 1956 Nobel Prize in chemistry. Most alkaloids taste bitter, and quinine is frequently used as a standard reference for bitterness in taste studies. The bitter taste of tonic water is due to quinine. Water solution of quinine salts are highly fluorescent, appearing light blue in

the presence of ultraviolet light (black light).

　[注]　invasion 侵略，Allied 連合国の，troops 軍隊，fluorescent 蛍光性の

[Ex.57][21] Connecting Biomass and Petroleum Processing with a Chemical Bridge

1. Petroleum is not only the primary feedstock for liquid fuels but also the basis for most of the chemicals and polymers that we use. Continuing questions about the longevity and stability of petroleum supplies, as well as the environmental impacts of its production and use, have driven the development of alternatives such as agricultural or woody biomass. If carbohydrates could be converted to compounds with fewer oxygenated groups that reacted more like petroleum, they could supply the petrochemical industry plants with a renewable feedstock.

　[注]　feedstock 原料，longevity 長寿命，alternatives 代替品，renewable 再生可能な，
　　　　not only —, but (also) 〜 : —だけでなく，〜も

2. Two recent papers by Bond *et al.* (1) and Lange *et al.* (2) now suggest that levulinic acid (LA), a dehydration product of simple sugars like glucose, can meet this need. Incorporation of LA as an intermediate allows use of catalytic conversion processes fully compatible with the infrastructure of the chemical industry.

　　　(1) J. Q. Bond, D. M. Alonso, D. Wang, R. M. West, J. A. Dumesic, *Science*, **327**, 1110 (2010).
　　　(2) J.-P. Lange *et al.*, *Angew. Chem. Int. Ed.*, **49**, 4479 (2010).

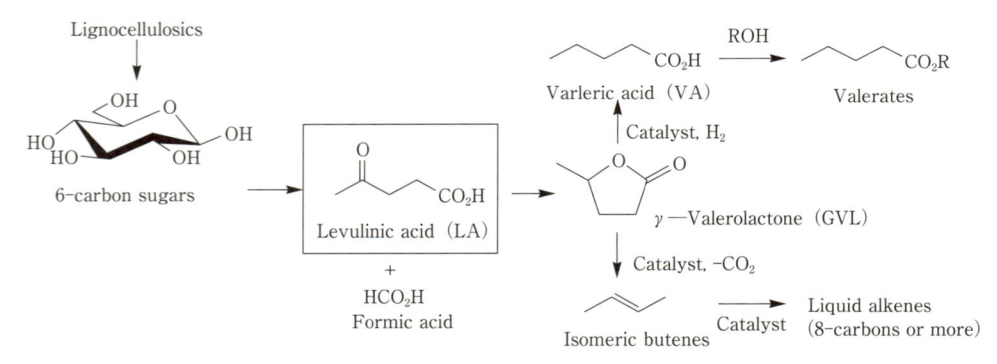

　[注]　infrastructure 構造基盤

3. Sugar dehydration, which is performed by treatment with acid, ultimately forms LA and formic acid in an approximately 3 : 1 weight ratio. This transformation has been known for decades, but has normally been accompanied by the formation of intractable by-products, and it is difficult to separate LA from the mixture. The status of LA as a biorefinery platform chemical was substantially upgraded in the early 1990s by Biofine Renewables and the development of a two-reactor system

that enabled the production of LA in high yield.

［注］　transformation 変換，accompany 伴う，intractable 扱いにくい，biorefinery platform バイオリファイナリー・プラットフォーム（生物資源変換のための施設），chemical 化学薬品，upgrade 格上げする

4. The conversion of LA to fuels begins with the well-established hydrogenation of LA to γ-valerolactone (GVL). Lange et al. optimized this conversion by testing more than 50 hydrogenation catalysts. Reduction of LA with 40-bar H_2 at 200℃, in the presence of a catalyst (1% by weight platinum metal dispersed on titanium oxide), proceeded with 95% conversion efficiency (the fraction of starting material that reacted).

［注］　optimize 最適化する，disperse 分散させる

5. Bond et al. describe a two-step process that catalytically eliminate CO_2 from GVL in aqueous solution, giving an initial mixture of isomeric butenes. Two or more butene units oligomerize to create higher hydrocarbons (eight or more carbon atoms) of sufficient molecular weight to serve as transportation fuels.

［注］　eliminate 脱離する，oligomerize オリゴマー化する（いくつかの単位が繋がる），transportation 輸送

6. Lange et al. hydrogenated LA to form valeric acid (VA) as the main product, which is then reacted with alcohols to form to a family of valerate esters suitable as gasoline- or diesel-fuel additives. The lower molecular weight esters (methyl, ethyl, and propyl valerate) exhibited performance suitable for use as a gasoline additive at levels of 10 and 20% (by volume). Higher esters (butyl and pentyl valerate) could be used directly as a diesel fuel or a diesel additive. Road trials using a 15% blend of ethyl valerate in regular gasoline were carried out with 10 vehicles and a total of about 250,000 km of driving. No engine performance issues were noted, but the lower energy density of ethyl valerate resulted in an expected loss of fuel economy per volume.

［注］　additive 添加物，performance 性能，vehicle 車

7. Thus, functionally useful biorefinery intermediates need not be structurally identical to compounds currently used in the petrochemical industry. Technology development issues must be addressed before a deeply entrenched and established biofuel (ethanol) can be replaced with promising but less recognized alternatives. Research challenges also exist: the initial LA production from sugars is still not highly efficient and has not been tested at scales necessary to ensure viability in a

fuel-production scenario.

　［注］　address 処置する，entrench 立場を守る，強固にする，viability　生育可能，成熟

8. Finally, the question of what to do with the formic acid produced as an unavoidable LA by-product must be addressed. A similar situation exists within the biodiesel industry, which must deal with the formation of a glycerol co-product. Formic acid can be used to reduce LA to GVL via catalytic transfer hydrogenation, but such processes have yet to be optimized. Nonetheless, these demonstrations of a biobased chemical as an intermediate in a process compatible with petrochemical technology will spur the use of renewable biomass as a source of the next generation of biofuels, and as a replacement for traditional raw material supplies.

　［注］　what to do with～：～をどうするか，compatible with～：～と両立できる，spur 拍車をかける

[Ex. 58][10] Amino Acids to Proteins

1. Proteins are large biomolecules that occur in every living organism. They are of many types and have many biological functions. The keratin of skin and fingernails, the insulin that regulates glucose metabolism in the body, and the DNA polymerase that catalyzes the synthesis of DNA in cells are all proteins. Regardless of their appearances or function, all proteins are chemically similar. All are made up of many amino acid units linked together by amide bonds in a long chain.

　［注］　polymerase ポリメラーゼ(重合を起こす酵素)，regardless of～：～に関わらず

2. Amino acids, as their name implies, are difunctional. They contain both a basic amino group and an acidic carboxyl group. Their value as building blocks for protein stems from the fact that amino acids can link together into long chains by forming amide bonds between the -NH$_2$ of one amino acid and the -CO$_2$H of another. For classification purposes, chains with fewer than 50 amino acids are usually called peptides, while the term protein is used for longer chains.

　［注］　imply 意味する，difunctional 2官能性の，stem from から生じる，classification 分類，while 一方で

3. Since amino acids contain both an acidic and a basic group, they undergo an intramolecular acid-base reaction and exist primarily in the form of a dipolar ion or zwitterion (German zwitter, meaning "hybrid"):

Amino acid zwitterions are salts and therefore have many of the physical properties associated with salts. They are soluble in water but insoluble in hydrocarbons and are crystalline substances with high melting points. In addition, amino acids are amphoteric : they can react either as acids or as bases, depending on the circumstances. In aqueous acid solution, an amino acid zwitterion is a base that accepts a proton to yield a cation ; in aqueous base solution, the zwitterion is an acid that loses a proton to form an anion.

［注］ intramolecular 分子内の，cf. intermolecular 分子間の，dipolar 双極性の，zwitterion ツヴィッターイオン（双性イオン），in addition 加えて，associated with 〜：〜と関連した，depending on〜：〜に依存して，amphoteric 両性の，either A or B：A または B（どちらか，どちらでも）．

[Ex. 59]$^{10)}$ Lipids

1. Lipids are small naturally occurring molecules that have limited solubility in water and can be isolated from organisms by extraction with a nonpolar organic solvent. Fats, oils, waxes, many vitamins and hormone, and most nonprotein cell-membrane components are examples. Note that this definition differs from the sort used for carbohydrates and proteins in that lipids are defined by a physical property (solubility) rather than by structure.

［注］ isolate 単離する，extraction 抽出

2. Lipids are classified into two general types: those like fats and waxes, which contain ester linkages and can be hydrolyzed, and those like cholesterol and other steroids, which don't have ester linkages and can't be hydrolyzed. Beeswax, for example, contains a lipid with the structure with an ester bond $CH_3(CH_2)_2CO_2(CH_2)_{27}CH_3$, which on hydrolysis will produce corresponding acid and alcohol : $CH_3(CH_2)_2CO_2H$ and $HO(CH_2)_{27}CH_3$.

[注]　classify into～：～に分類する，corresponding (*adj*) 対応する，produce 生産する，生じる

3. Animal fats and vegetable oils are the most widely occurring lipids. Although they appear different — animal fats like butter and lard are solids, whereas vegetable oils like corn oil and peanut oil are liquids — their structures are closely related. Chemically, fats and oils are triacylglycerols (also called triglycosides), triesters of glycerol with three long chain carboxylic acids. Hydrolysis of a fat or oil with aqueous NaOH yields glycerol and sodium salts of three long-chain fatty acids:

$$\begin{array}{c} CH_2O-COR \\ | \\ CHO-COR' \\ | \\ CH_2O-COR'' \end{array} \xrightarrow{aq\ NaOH} \begin{array}{c} CH_2OH \\ | \\ CHOH \\ | \\ CH_2OH \end{array} + \begin{array}{c} RCOONa \\ R'COONa \\ R''COONa \end{array}$$

Fat or Oil　　　　　　　　Glycerol　　　Fatty acid, Na salt
　　　　　　　　　　　　　　　　　　　　　　(soap)

[注]　yield 生じる

4. The fatty acids obtained are generally unbranched and contain an even number of carbon atoms between 12 and 20. If one or more double bonds are present, they usually have *Z* (*cis*) geometry. The three fatty acids of a specific molecule need not be the same, and a fat or oil from a given source is likely to be a complex mixture of many different triacylglycerols.

[注]　unbranched 分岐のない(直鎖の)，even 偶数の *cf.* odd 奇数の，geometry 幾何学, 幾何異性(*cis-trans* または *Z-E* 異性)，likely to～：～である可能性が高い，～らしい

5. More than 100 different fatty acids are known, and about 40 occur widely. Palmitic acid (C_{16}) and stearic acid (C_{18}) are the most abundant saturated fatty acids, while oleic, linoleic, and linolenic acids (all C_{18}) are the most abundant unsaturated ones. Oleic acid is monounsaturated because it has only one double bond, whereas linoleic and linolenic acids are polyunsaturated because they have more than one double bonds; two and three double bonds, respectively.

[注]　abundant 豊富にある，saturated 飽和の，unsaturated 不飽和の，while (whereas) 一方で

6. Soap has been known since at least 600 BC, when the Phoenicians prepared a curdy material by boiling goat fat with extracts of wood ash. The cleaning properties of soap weren't generally recognized, however, and the use of soap don't become widespread until the 18th century. Chemically, soap is a mixture of the sodium or potassium salts of long-chain fatty acids produced by hydrolysis (saponification) of

fats with alkali as shown above.
　［注］　curdy 凝乳状の，saponification けん化（アルカリ加水分解）

7. Crude soap curds contain glycerol and excess alkali as well as soap but can be purified by boiling with water and adding NaCl to precipitate the pure sodium carboxylate salts. The smooth soap that results is dried, perfumed, and pressed into bars. Dyes are added for colored soaps, antiseptics are added for medicated soaps, pumice is added for scouring soaps, and air is blown in for soaps that float.
　［注］　curd 凝乳（カード），perfume 香料をつける，antiseptic 防腐剤，pumice 軽石，scour 磨く

8. Soaps act as cleansers because the two ends of a soap molecule are so different. The carboxylate end of the long-chain molecule is ionic and therefore hydrophilic (water-loving). As a result, it tries to dissolve in water. The long aliphatic chain portion of the molecule, however, is nonpolar and hydrophobic (water-fearing). It tries to avoid water and to dissolve in grease. The net effect of these two opposing tendencies is that soaps are attracted to both grease and water and are therefore useful as cleansers.
　［注］　cleanser クレンザー（洗剤），hydrophilic 親水性の，hydrophobic 疎水性の

9. When soap molecules are dispersed or dissolved in water, the long hydrocarbon tails cluster together into a hydrophobic ball, while the ionic heads on the surface stick out into the water phase. These spherical clusters, called micelles, are shown schematically below. Grease or oil droplets are solubilized in water when they become coated by the hydrophobic nonpolar tails of soap molecules in the center of micelles. Once solubilized, the grease and dirt can be rinsed away.

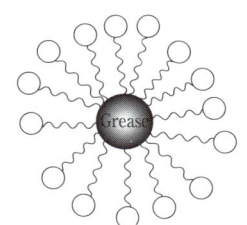

A schematic model of a soap micelle solubilizing grease or dirt in the core

　［注］　cluster 群がる，solubilize 可溶化する，rinse すすぐ

10. Soaps make life much more pleasant than it would otherwise be, but they also have drawbacks. In hard water, which contains metal ions, soluble sodium carboxylates are converted into insoluble calcium and magnesium salts, leaving the scum around bathtubs and the gray tinge on clothes. Chemists have circumvented

these problems by synthesizing a class of synthetic detergents based on salts of long-chain alkylbenzenesulfonic acids. Unlike soaps, sulfonate detergents don't form insoluble metal salts in hard water and don't leave an unpleasant scum.

Sodium alkylbenzenesulfonate
(R = mixture of C12 alipahatic chains)

［注］ scum かす, tinge 薄い色, circumvent 免れる

11. Steroids are plant and animal lipids with a characteristic four-ring carbon skeleton, as shown below for a choresterol. Steroids occur widely in body tissue and have many different kinds of physiological activity. Among the more important kinds of steroids are the cholesterols, the sex hormones (androgens and estrogens), and the adrenocortical hormones.

Cholesterol

［注］ adrenocortical 副腎皮質の

[Ex. 60] 15) Digestion of Carbohydrates

1. Carbohydrates are a class of compounds that includes polyhydroxy aldehydes, polyhydroxy ketones, and large molecules that can be broken down to form polyhydroxy units upon hydrolysis. Monosaccharides cannot be broken down into smaller units. Monosaccharides, or simple sugars, are classified by the number of carbon atoms in the molecule and also by the kind of carbonyl functional group found in the structure. Thus, glucose is an aldohexose (an aldehyde containing six carbon atoms), and fructose is a ketohexose (a ketone with six carbon atoms).

2. Before the cells of your body can utilize the energy stored in carbohydrates present in your diet, the carbohydrates must be digested and absorbed. Digestion is the process by which complex molecules are broken into simple molecules. These simple molecules pass through the intestinal wall into the bloodstream during absorption. Absorption is a form of dialysis, i. e., the movement of small molecules through a membrane.

　　［注］ digest 消化する, absorb 吸収する, digestion 消化, absorbtion 吸収, intestinal 腸の, dialysis 透析

3. The digestion of carbohydrates begins in the mouth as teeth tear food into tiny pieces; smaller pieces have a greater surface area and will be digested faster. Saliva contains an enzyme (amylase) that begins the hydrolysis of starch (a large polysaccharide) to dextrins (small polysaccharides) and maltose (a disaccharide). After swallowing the food enters the stomach where protein and fat digestion begin but carbohydrate digestion temporalily ceases; the low pH of the stomach's gastric juice inactivates the salivary enzymes.

　　［注］　saliva 唾液，polysaccharide 多糖，disaccharide 二糖，swallow 飲み込む，gastric juice 胃液

4. As food passes into the small intestine it is neutralized by alkaline pancreatic and intestinal juices. Those juices also contain an enzyme that renews the hydrolysis of complex carbohydrates. Eventually all polysaccharides and disaccharides are broken down to glucose, fructose, and galactose. These monosaccharides are small enough in size to pass through the intestinal wall and are absorbed into the blood. After circulating in the blood, fructose and galactose are converted into glucose by the liver. The glucose in the blood may be immediately used to provide energy for cellular activities or it may be stored as glycogen (a polysaccharide) in the liver and the muscles.

　　［注］　small intestine 小腸，pancreatic 膵臓の

5. An aqueous solution of glucose is an equilibrium between three forms : α-glucose, β-glucose and open-chain glucose. The open-chain structure makes up only 0.02 percent of the equilibrium mixture. Aqueous glucose exists 36 percent in the α-structure and 64 percent in the β-structure. α- and β-Gucose are cyclic structures in which five carbon atoms and one oxygen atom form a six-sided ring. Six-sided glucose rings are not flat but exist in a puckered "chair" conformation. The only difference between the α-and β-forms is the position of one hydroxyl group on carbon-1. If the hydroxyl group is positioned pointing down (axial), it is the α-form. If the hydroxy group is up (equatorial), it is the β-form.

　　［注］　open-chain 開鎖（環が開いて鎖状の形）の，six-sided 六辺の，puckered ひだになった，"chair" conformation "いす形" 配座

α-D-glucose ⇌ open-chain D-glucose ⇌ β-D-glucose

[Ex. 61] [10] Nucleic Acids and Heredity

1. The genetic information of an organism is stored as a sequence of deoxyribonucleotides strung together in the DNA chain. For the information to be preserved and passed on to future generations, a mechanism must exist for copying DNA. For the information to be used, a mechanism must exist for decoding the DNA message and implementing the instructions it contains.

[注] sequence 連鎖, strung (string 糸になる) の過去分詞, for～ to —: ～が—するために, preserve 保存する, decode 暗号を解く, 解読する, implement 実行する

2. What Crick called the "central dogma of molecular genetics" says that the function of DNA is to store information and pass it on to RNA. The function of RNA, in turn, is to read, decode, and use the information received from DNA to make proteins. By decoding the right bit of DNA at the right time, an organism uses genetic information to synthesize the thousands of proteins necessary for functioning.

Replication ↻ DNA —Transcription→ RNA —Translation→ Proteins

[注] dogma 教義, 定説, in turn 引き続いて, 順繰りに

3. Three fundamental processes take place in the transfer of genetic information: *Replication* is the process by which identical copies of DNA are made so that genetic information can be preserved and handed to succeeding generations. *Transcription* is the process by which the genetic messages are read and carried out of the cell nucleus to ribosomes, where protein synthesis occurs. *Translation* is the process by which the genetic messages are decoded and used to synthesize proteins.

[注] replication 複写, succeeding 続く, 次の, transcription 転写, translation 翻訳, so that ～: ～するように, (結果)～となる

[Ex. 62]¹²⁾ Polyethylene and Polypropylene

1. Ethylene and propylene, among many other molecules with double bonds, form what are called "addition polymers." This means that no atoms of the polymerizable alkene are lost on formation of the polymer. Many thousands of ethylene and propylene monomers may bond together to form polyethylene and polypropylene, respectively.

［注］ addition polymers 付加重合体（付加重合で生成するポリマー）

2. Laying aside the mechanism of the addition, the overall process involves converting the alkene π-bonds to the σ-bonds holding the repeating units together in the polymer. Since the σ-bonds are far stronger than the disrupted π-bonds of the monomers, a great deal of energy is released and the thermodynamic picture involves a large negative enthalpy, ΔH, favoring the polymerization, which is an exothermic reaction.

［注］ repeating units 反復（繰り返し）単位，exothermic 発熱の

3. On the other hand, the holding together in the polymer chain of the many alkene molecules that had been free to move independently causes a large reduction in entropy for the process, so that ΔS is negative, disfavoring the polymerization. The polymerization is an ordering process.

［注］ ordering 秩序化の

4. These two competing thermodynamic factors control all polymerizations, and with ethylene and propylene, at the temperature of the polymerization, the enthalpic term overwhelms the entropic term. If this were not the case, a polymer could not be formed, and in fact because of these competing enthalpy and entropy factors all polymers can only be formed below a certain temperature known as the *ceiling temperature*.

［注］ *ceiling temperature* 天井温度（この温度を超えると重合が起こらない）
付注：エチレンの重合の熱力学について解説
　　　C＝C（$\sigma + \pi$）結合解離エネルギー ＝ 420 kJ mol⁻¹
　　　C－C（σ）　結合解離エネルギー ＝ 250 kJ mol⁻¹
　　　エチレンの重合では，エチレン分子当たり，1つの二重結合が切れて，2つの一重結合が生成する（または，1つのπ結合が切れて，1つのσ結合になる）ことになるので，
　　　n CH₂＝CH₂ ⟶ －[CH₂－CH₂－]ₙ－ ： ΔH ＝ 420 － 2×250 ＝ － 80 kJ mol⁻¹
　　　　　　　　　　　　　　　　　　　　　（または，170 － 250 ＝ －80）
　　　一方，　　ΔS＝ -150 J K⁻¹ mol⁻¹　（Polymer Handbook, 4th ed., 1999 から）
　　　∴重合のギブス自由エネルギー変化： ΔG（J mol⁻¹）＝ $\Delta H - T \Delta S$ ＝ －80,000 ＋ 150T

天井温度(T_c)では$\Delta G = 0$なので，$T_c = 80,000/150 = 533$ K $= 260$℃
この天井温度以下で$\Delta G < 0$（重合可能），天井温度以上で$\Delta G > 0$（重合不可能：解重合（モノマーに戻る））

5. While the addition of a source of free radicals to a gaseous sample of ethylene rapidly produced a white powder, polyethylene, which can be melted and molded into many of its familiar uses, the addition of such a radical to propylene produced a useless paste of short hydrocarbon molecules. Thus, the chain growth of propylene was stopped before it could get going, in spite of the fact that the thermodynamic picture outlined as above for ethylene also fits propylene: there is no impediment from thermodynamic consequences for the formation of polypropylene. This fact has been reasonably accounted for by the blocking role of allyl-resonance ; that is, the possible formation of the resonance-stabilized allyl radical from propylene stands in the way of free radical polymerization.

[注] mold 成形する，impediment 障害，allyl-resonance アリル共鳴，consequence（必然の）結果，stand in the way of 立ちはだかる，

付注：プロピレンのラジカル重合は，メチル基からの水素引き抜き連鎖移動で生じるアリルラジカルが共鳴安定化するために，重合の進行が妨げられる。

$R\cdot + CH_2=CH\text{-}CH_3 \longrightarrow R\text{-}H +\quad CH_2=CH-CH_2\cdot$
$\qquad\qquad\qquad\qquad\qquad\qquad\qquad\qquad\updownarrow$ "アリル共鳴"
$\qquad\qquad\qquad\qquad\qquad\qquad\qquad\cdot CH_2 - CH = CH_2 \cdot$

一方，現在，Ziegler-Natta 触媒やメタロセン触媒による配位アニオン重合によって，立体規則性のポリプロピレンが大量に生産されている。Ex. 71 参照。

[Ex. 63][12)]Rubber Elasticity

As with many polymers, the polymer chains in natural rubber are very long and exist in the relaxed state in a randomly coiled shape. The best way to imagine a random coil is to think of tracing a random path in three dimensions. This will be a very disorderly array, which has been compared to a plate of cooked spaghetti. When the rubber is stretched, the individual polymer chains are forced to change to a much more extended shape. Upon removing the stress, the chains immediately relax to their original, random coils. For a polymer to be an elastomer it must have a structure that allows facile and reversible change of its shape with little change of energy. Thus, rubber elasticity essentially results from conformational change, and is a so-called entropic elasticity, since the stretching and the relaxation involve the loss and the gain in entropy, respectively.

[注] elasticity 弾性，as with と同様に，relax 緩む，disorderly 無秩序な，array 並び，compared to に比較される（同じようである），stretch 延伸する，force to（強制的に）させる，immediately 直ちに，elastomer 弾性体（ゴム），respectively それぞれ

[Ex. 64][22] Silicones

1. Silicones are polymers whose backbones are long, flexible chains of alternating silicon and oxygen atoms. Dangling from the backbone like charms from a bracelet are side chains, usually small, carbon-based units, and the choice of these side chains gives silicones a remarkable range of properties. The water-repellent grease has oily, nonpolar side chains such as methyl group. The nonpolar side chains and the polar water molecules do not mix, repelling the water from the silicone.

[注]　silicone シリコーン（— SiR_2-O —を基本構造とするポリマー：Si 置換基の R が側鎖）, silicon シリコン（珪素）, alternating 交互の, dangle ぶら下がる, charm（腕輪の）飾り

2. Another application of silicones depends on a careful balance between polar and nonpolar side chains. Small amounts of silicone foaming agent control the bubble size in polyurethane foams. A high proportion of polar side chains makes the foam foamier. The bubbles become bigger, forming open pores and producing the soft foams found in car seats and furniture cushions. Reduce the number of polar side chains, and the bubbles remain small. These tiny bubbles do not open up to form pores, and the foam is a much stiffer solid used for insulation.

[注]　foaming agent 発泡剤, insulation 絶縁

3. Silicones have other, seemingly contradictory properties. A silicone resin coating the bread pans in a bakery keeps fresh-baked bread from sticking in the pan, and a liquid silicone polymer on the molds in the factories does the same thing for newly made tires. But adding a "tackifying resin" makes the silicone sticky and produces the drug-permeable contact adhesive used on those skin patches containing nicotine (for smokers who are trying to quit) or scopolamine (for seasickness sufferers who are trying not to lose it).

[注]　seemingly 見かけは, contradictory 矛盾した, keep — from 〜：—を〜から防ぐ, mold 鋳型, tackify 粘着性を増す, patch パッチ（布切れ）

4. Silicon and oxygen are two most abundant elements on Earth, and they combine naturally to form silicates, including glass and such materials as quartz and granite. These two elements were first combined synthetically — as silicones — in the United States in the 1930s. They were originally expensive and unhandy to make, but the discovery of a cheaper, easier method of producing them, coinciding with the interest in their novel properties sparked by World War II, started an avalanche of research into new uses for these versatile polymers that continue unabated today.

[注]　granite 花崗岩, coincide with と一致する, spark 引き起こす, avalanche なだれ, 殺到, versatile 多芸の, unabated 減じない, 変わらない

[Ex. 65][22] Implanted Polymers for Drug Delivery

1. We have all heard that biodegradable polymers are good for the environment. But they may be good for cancer patients, too. Efforts are now under way to design polymer implants that will slowly degrade inside the human body, releasing cancer-fighting drugs in the process.

[注] biodegradable 生分解性の, under way 進行中の, implant (n) インプラント(体内に埋め込まれる器具, 材料), (vt) インプラントする(体内に埋め込む), degrade 分解する, release 放出する

2. Such an implant would need several specific properties. It would have to degrade slowly, from its outside surface inward, so that a drug contained throughout the implant would be released in a controlled fashion over time. The polymer as a whole should repel water, protecting the drug within it — as well as the interior of the implant itself — from dissolving prematurely. But the links between the monomers — the building blocks that make up the polymer — should be water-sensitive so that they will slowly fall apart.

[注] repel 反発する, はじく, prematurely 早すぎて, fall 崩壊する

3. Anhydride linkages — formed when two carboxylic-acid-containing molecules join together into a single molecule, creating and expelling a water molecule in the process — are promising candidates, because water molecules rapidly split the anhydride linkages in the reverse of the process that created them, yet the polymer molecules can still be water-repellent in bulk. By varying the ratios of the components, surface-eroding polymers lasting from one week to several years have been synthesized.

[注] anhydride (酸)無水物, surface-eroding 表面侵食する

4. These polymer disks are now being used experimentally as a postoperative treatment for brain cancer. The surgeon implants several polyanhydride disks, each about the size of a quarter, in the same operation in which the brain tumor is removed. The disk contains powerful cell-killing drugs called nitrosoureas. Nitrosoureas are normally given intravenously, but they are indiscriminately toxic, and this approach generally damages other organs in the body while killing the cancer cells. But placing drug in the polymer protects the drug from the body, and the body from the drug. The nitrosoureas lasts approximately the duration of the polymer — in this case, nearly one month. And the eroding disk delivers the drug only to the immediate surroundings, where the cancer cells lurk. The polymer degradation method of drug delivery is making good progress toward approval by the Food and Drug Administration.

[注] postoperative treatment 手術後治療, quarter 1/4, 25セント銀貨, intravenously 静

脈内で，indiscriminately 無差別に，duration 持続期間，lurk 潜む

[Ex. 66][22] Biopolymers versus Synthetic Polymers

1. The volume of biopolymers in the world far exceeds that of synthetic macromolecules. Biological polymers include DNA, RNA, proteins, carbohydrates, and lipids. DNA and RNA are informational polymers (encoding biological information), while globular proteins, some RNAs, and carbohydrates serve chemical functions and structural purposes. In contrast, most synthetic polymers, and fibrous proteins such as collagen (which makes up tendon and bone) and keratin (which makes up hair, nails, and feathers), are structural rather than informational or chemically functional. Structural materials are useful because of their mechanical strength, rigidity, or molecular size, properties that depend on molecular weight, distribution, and monomer type.

［注］ biopolymer 生体高分子，macromolecule 巨大分子，高分子，informational 情報の，encode 符号（暗号）化する，in contrast 対照的に，globular 球状，make up 成す，tendon 腱

2. In contrast, informational molecules derive their main properties not simply from their size, but from their ability to encode information and function. They are chains of specific sequences of different monomers. For DNA the monomers are the deoxyribonucleic acid bases; for RNA, the ribonucleic acid bases; for proteins, the amino acids; and for carbohydrates or polysaccharides, the sugars. The paradigm in biopolymers is that the sequence of monomers along the chain encodes the information that controls the structure or conformation of the molecule, and the structure encodes the function. An informational polymer is like a necklace, and the monomers are like the beads.

［注］ paradigm 範例，sequence 順序，配列

3. For RNA and DNA, there are 4 different monomers (beads of different colors). Information is encoded in the sequence of bead colors, which in turn controls the sequence of amino acids in proteins. There are 20 different colors of beads. A globular protein folds into one specific compact structure, depending on the amino acid sequence. This balled-up shape, or structure, is what determines the protein functions. The folding of the linear structure produces a three-dimensional shape that controls the function of the protein through shape selection.

［注］ in turn 引き継いで，順繰りに

[Ex. 67][22] Films, Membranes, and Coatings

1. Polymers are used in many applications in which their main function is to regulate the migration of small molecules or ions from one region to another. Examples include containers whose walls must keep oxygen outside or carbon dioxide and water inside; coatings that protect substrates from water, oxygen, and salts; packaging films to protect foodstuffs from contamination, oxidation, or dehydration ; so-called "smart packages," which allow vegetables to respire by balancing both oxygen and carbon dioxide transmission so that they remain fresh for long storage or shipping times ; thin films for controlled delivery of drugs, fertilizers, herbicides, and so on; and ultrathin membranes for separation of fluid mixtures.

[注] migration 移動,substrate 基質,contamination 汚染,dehydration 脱水,respire 呼吸する,transmission 移送,shipping 輸送,fertilizer 肥料,herbicide 除草剤,ultrathin 超薄の

2. These diverse functions can be achieved partly because the permeability to small molecules via a solution-diffusion mechanism can be varied over enormous ranges by manipulation of the molecular and physical structure of the polymer. The polymer that has the lowest known permeability to gases is bone-dry poly (vinyl alcohol), while the recently discovered poly (trimethylsilyl propyne) is the most permeable polymer known to date. The span between these limits for oxygen gas is a factor of 10^{10}.

[注] diverse 多様な,permeability 透過性,solution-diffusion 溶解—拡散,manipulation 操作,bone-dry からからに乾燥した

3. A variety of factors, including free volume, intermolecular forces, chain stiffness, and mobility, act together to cause this enormous range of transport behavior. Recent experimental work has provided a great deal of insight, while attempts to simulate the diffusional process using molecular mechanics are at a very primitive stage. There is clearly a need for guidance in molecular design of polymers for each of the applications. In addition, innovations in processing are needed.

[注] intermolecular 分子間の,(cf intramolecular 分子内の),transport 輸送,primitive 初期の,innovation 技術革新,processing 加工

[Ex. 68][23] Room-Temperature Ionic Liquids

1. Room-Temperature Ionic Liquids (RTILs) present many opportunities to reassess and optimize existing technologies and processes. RTILs are classified as molten (typically organic) salts at ambient conditions. These intrinsic properties of RTILs that differentiate them from common organic solvents and water are: nonvolatility,

thermal stability and tunable chemistry. These properties can impart significant advantages to RTILs for chemical engineering applications, especially in gas separations and enhanced gas solubility.

 [注] ionic liquid イオン液体，reassess 再評価する，optimize 最適化する，existing 既存の，ambient 周囲の，intrinsic 固有の，differentiate A from B：A を B と区別する，nonvolatility 不揮発性，tunable 調和可能な，impart 付与する，significant 意味のある，重要な

2. RTILs present a highly versatile and tunable platform for the development of new materials (solvents, polymers, gels) aimed at the capture of CO_2. The solubility of CO_2 is higher in RTILs compared to common solvents. The addition of specific complexing agent, such as amine, that can reversibly bind with CO_2 provides the capability to increase the CO_2 solubility by a factor of 10. The tunable chemistry of imidazolium-based RTILs also allows for the generation of new types of CO_2-selective polymer membranes.

$$H_3C-N\overset{\oplus}{\underset{\ominus NTf_2}{\frown}}N-R \quad (Tf=OSO_2CF_3)$$

 [注] versatile 多芸多才の，platform 演台，舞台，capture 捕獲，complexing agent 錯化剤，by a factor of～：～倍で，CO_2-selective　CO_2 選択性の

3. Given their CO_2-selective properties, there has been a great deal of interest in using RTILs as the selective component in membranes. A straightforward approach to use RTILs in a membrane configuration is to employ supported ionic liquid membranes. In general, supported liquid membranes are composed of a liquid immobilized within a polymer or inorganic support. They typically provide larger gas permeabilities than conventional polymer membranes, as gas diffusion through a dense liquid film is often much more rapid than through a rubbery or glassy polymer.

 [注] membrane 膜，supported 担持（支持）された，support 支持体，担持体，immobilize 固定化する，permeability 透過性

4. However, traditional supported liquid membranes suffer from issues relating to evaporative losses of the liquid phase into the gas stream, resulting in degradations of membrane integrity and performance. Supported RTIL-membranes circumvent this limitation, as the RTIL component cannot evaporate. Evaluation of several different imidazolium-based RTILs revealed that the supported RTILs possessed permeability and selectivity properties for CO_2/N_2 that were superior to most conventional polymer membranes.

 [注] traditional, conventional 伝統的な，従来の，suffer from を苦しむ，degradation 分解，低下，integrity 完全，本来の形，circumvent 逃れる，result in～：（結果）～とな

る，〜を生じる

[Ex. 69][24] Block Copolymer Micelles

1. The ability of block copolymers to self-assemble into micellar aggregates in a selective solvent is of potential interest in a variety of technologies, including, for example, drug delivery and therapeutics, catalysis, separations, cosmetics, and food science. Although in many respects analogous to lipids, surfactants, and small molecule amphiphiles, block copolymers offer many advantages in terms of tunability of properties.

　［注］　block copolymer ブロック共重合体，self-assemble 自己集合する，aggregate 凝集体，selective 選択的，therapeutics 治療(学)，amphiphile 両親媒種

2. For a given AB diblock copolymer in a solvent that is good for A, the canonical micelle shapes are sphere, worm, and vesicle. The factors that govern the choice of shape under equilibrium conditions are rather well-known; for example, increasing interfacial tension between B and the solvent drives the transition to larger micelles and flatter interfaces, whereas crowding of the well-solvated A corona blocks acts in the opposite sense.

　［注］　canonical 標準の，worm 虫(のような形)，interfacial tension 界面張力，well-solvated よく溶媒和された

付注：(1) AB ジブロック共重合体：A と B 2種のモノマーが AAA…A — BBB…B のように A 鎖と B 鎖が連続的に結合した共重合体。
　　cf. ランダムまたは統計的共重合体(random or statistical copolymer)：A と B がランダムまたは統計的に結合した共重合体。
　　グラフト共重合体(graft copolymer)：A 鎖と B 鎖が幹と枝の分岐状に結合した共重合体。
(2) A のコロナ：AB ジブロック共重合体で，A の良溶媒中では，A 鎖がミセルの外側に放射状(コロナ状)に突き出る。

[Ex.70][20] Serendipity

One way in which a person can prepare to benefit from serendipity is through careful and intensive study in the field of chosen investigation. The American physicist Joseph Henry paraphrased Pasteur's dictum when he said, "The seeds of great discovery are constantly floating around us, but they only take root in minds well prepared to receive them." For example, although Fleming was not looking for an antibacterial agent at the time a spore floated into his petri dish, he was extremely well read and trained in microbiology and could readily recognize the meaning of the clear area in the bacterial culture produced by the accidental implantation of the mold.

［注］ benefit 恩恵を受ける，paraphrase 意訳する，dictim 格言，antibacterial agent 抗菌剤，serendipity セレンディピティ（予期しない偶然の発見），petri dish ペトリ皿（シャーレ），spore 胞子

[Ex. 71] [12)] Serendipity in Polyethylene and Polypropylene

1. Polyethylene and polypropyrene are not only huge molecules, but they are also formed and used in our world in huge amounts. The amount of polyethylenes produced worldwide each year can be measured in the hundreds of billions of kirograms, and they find use in numerous applications. Morawetz recounts in interesting detail the discovery of polyethylene. The discovery arose from research conducted at ICI Corporation in Englnd in the years before World War II as a consequence of a decision to explore chemical reactions under high pressure. ICI saw this as a foray into basic research and it is interesting that at this time in the late 1920s DuPont made a similar decision and hired Wallace Carothers to carry out research also without aiming at commercial products. DuPont's decision led to the invention of nylon, while that of ICI led to polyethylene. Not a bad idea to carry out basic research!

［注］ recount 詳説する，consequence 結果，foray 侵略，介入，hire 雇う

2. The polymerization of ethylene occurred accidentally at ICI in a misguided attempt to add ethylene to benzaldehyde under high pressure. The benzaldehyde was recovered unchanged, but a waxy solid was obtained that proved to be a relatively low molecular-weight polymer of ethylene. After several experiments to try to improve the result − and some delay caused by explosion − it was realized that a high molecular-weight polyethylene, which is a solid material in contrast to the waxy substance originally obtained, could be made using a high pressure of ethylene with a small concentration of oxygen. The insight about oxygen arose from another mistake − a leak in the apparatus which allowed air to enter inadvertently.

［注］ explosion 爆発，insight 洞察，inadvertently 不注意で

3. We do not intend to mislead our readers to believe that experiments conducted in a sloppy manner, and aiming for impossible goals are necessary for success in chemistry. But in fact we can see accidental discovery again in the work that lead to the Nobel Prize for Ziegler and Natta. These serendipious discoveries were only possible because these chemists, who were very much innovators and skilled in their profession, were also prepared to fully understand the surprising implications of the results of their experiments. One is reminded of a famous quote of Louis Pasteur: "In the field of observation, chance favors only the mind that is prepared."

[注] sloppy ずさんな，innovator 革新家，be prepared to 準備ができている

4. Stereospecific polymerization of propylene also actually begins with an accidental discovery of a new way to polymerize ethylene. This arose from contamination of a chemical apparatus in the laboratory of Karl Ziegler, a distinguished German professor who specialized in catalysis. In a series of experiments in Ziegler's laboratory which started just after World War II, and was intended to polymerize ethylene by methods other than those used in the ICI work, it was discovered that one of the autoclaves used for the experiments was especially effective in producing butene in high yield. This was traced eventually to the presence of colloidal nickel that had been produced in the cleaning of the stainless autoclave and to subsequent exposure to lithium used in the experiments.

[注] stereospecific 立体特異性の，contamination 汚れ，apparatus 装置，distinguished 著名な，other than と別の

5. The result stimulated an exploration across the spectrum of metals in the periodic table to find others that would optimize this result. In one of these ironies of scientific work the testing of many metals was not successful in achieving the objective of producing butene from ethylene but rather led to success in achieving the original objective of polymerizing ethylene. The addition of some metals, zirconium and titanium, transformed ethylene to high molecula-weight-polyethylene. And it was realized immediately that the polyethylene produced by this new method differed from that produced by the ICI high-pressure process.

[注] exploration 探索

6. In an early use of infrared spectroscopy, the new polymer seemed to have fewer pendant methyl groups than that produced by the ICI process, and it also softened at a higher temperature. From the work in Ziegler's laboratory, a catalyst had therefore been developed allowing a polyethylene to be produced that differed in an important way from the familiar polyethylene produced by the initiation step with free radicals.

[注] pendant ペンダント(垂れ下がった)，allow～to —：～を—させる，initiation step 開始段階

あとがき

There is no royal road to learning！
（学問に王道なし）

　もとより著者らは語学としての「英語」を生業とするものではないが，高分子の基礎研究に携わる中で，道具としての英語の恩恵を多く受けた。学生時代からの研究生活を通じて，英語で書かれたあまたの著書や論文は，多くの啓発・感銘・刺激を与えてくれた。研究室のゼミや「化学英語」の授業を担当した経験と合わせて，英語に限らないが，外国語から新しい知見・考えを経験する喜び・興奮をいささかでも若い読者に伝えることができれば，と期待したことがこの演習書を書くきっかけとなった。化学および関連の専門分野に進まれる方々や研究に携わっておられる卒研生や院生の方々に，何らかの参考になれば，という思いである。「和訳例」は，演習書として直訳・対訳を旨としたために，日本語の文章として若干ぎくしゃくとしたところがあるかもしれない。また，著者らの不注意・不勉強から，間違った解説，不適当な訳もあるかもしれない。読者諸氏の忌憚のないご意見を賜れば幸いである。

引用文献

1) Nivaldo J. Tro, "Chemistry in Focus: A Molecular View of Our World 2nd Ed.," Brooks/Cole, California, (2001)
2) 中村菀爾, 『化学英語演習』増補 3 版, 共立出版 (1998)
3) L. Pauling, "General Chemistry," Dover Publications (1970)
4) A. Serefani, "Linus Pauling, A Man and His Science", Paragon House, New York (1989)
5) O. Shimomura (下村 修), Nobel Lecture (Dec. 8, 2008). http://nobelprize.org/nobel_prizes/chemistry/laureates/
6) I. M. Klotz, R. M. Rosenberg, "Chemical Thermodynamics 5th Ed.", John Wiley & Sons (1994)
7) G. Rayner-Canham, T. Overton, "Descriptive Inorganic Chemistry", 4th Ed., W. H. Freeman & Co. (2006)
8) L. F. Fieser, and M. Fiezer, "Introduction to Organic Chemistry," D. C. Heath and Company, (1957) 丸善版から.
9) R. Noyori (野依良治), Nobel Lecture (Dec. 8, 2001). http://nobelprize.org/nobel_prizes/chemistry/laureates/
10) J. McMurry, E. Simanek, "Fundamentals of Organic Chemistry, 6th Ed.", Thomson Brooks/Cole, USA (2007)
11) ACS, "The Stellar Thread — A Story of DNA, Evolution, and the Immortality of Ideas," (1981)
12) M. M. Green, H. A. Wittcoff, "Organic Chemistry Principles and Industrial Practice," Wiley-VCH (2003)
13) http://nobelprize.org/nobel_prizes/chemistry/laureates/2010/press.html
14) http:/www.organic-chemistry.org/namedreactions/Suzuki-coupling.shtm
15) D. B. Macaulay, J. M. Bauer, M. M. Bloomfield, "General, Organic, and Biological Chemistry: An Integrated Approach", John Wiley & Sons (2007)
16) J. W. Moore, C. L. Staniski, P. C. Jurs, "Chemistry," Harcourt College Pub. (2002)
17) G. M. Lewis, M. Randall, Revised by K.S.Pitzer and L.Brewer, "Thermodynamics 2nd Edition", McGraw-Hill Book Company, New York (1961)
18) L. P. Hammett, "Physical Organic Chemistry" 2nd Ed., Mcgraw-Hill (1970)
19) K. Tsujii, "Surface Activity", Academic Press (1998)
20) R. M. Roberts, "Serendipity", John Wiley & Sons, Inc. (1989)
21) J. J. Bozell, *Science*, **329**, 522, July 30 (2010)

22) National Research Council, "Polymer Science and Technology", National Academy Press, Washington D.C (1994)
23) R. D. Noble, *Polym. Prepr., Japan,* **59**(1), 4 (2010)
24) T. P. Lodge, *Polym. Prepr., Japan,* **59**(1), 19 (2010)

その他，和訳，基礎編，訳注，解説の編集にあたっては，次の図書も参考にしたことを記して，感謝申し上げる。
(1) 岩波書店，『理化学辞典』，第4版 (1987)
(2) 千原秀昭ほか編，『化学英語の活用辞典』，化学同人 (1970)
(3) 飯田隆編，『ライフサイエンス系の基礎英語テクニカルターム』，三共出版 (2008)
(4) 多田旭男，上松敬禧，中平隆幸，中野勝之，『アクティブ科学英語』，三共出版 (1997)
(5) 『新英和大辞典』研究社 (1980)
(6) J. Brandrup, E. H. Immergut, E. A. Grulke, A. Abe, D. R. Bloch, "Polymer Handbook" 4th Ed., Wiley-Interscience (1999)
(7) 蒲池幹治，岩井薫，伊藤浩一，『基礎物質科学―大学の化学入門』，三共出版 (2007)

和 訳 例
(Ex. 25 ～ Ex. 71)

[Ex. 25] 化学と社会

1. 化学は，巨視的現象に対する分子的な理由を調べる科学である。それは，観察と実験を強調する科学的な方法を使って，我々の毎日の世界と分子・原子の世界の間の繋がりを探求する。

2. 紀元前600年頃，人々は，宇宙世界とその挙動の基礎をなす道理について知りたいと思った。何人かのギリシャの哲学者，プラトン，デモクリトス，ターレス，エンペドクレス，アリストテレスらは，道理だけで自然の神秘を解決できると信じた。彼らは，自然界の我々の理解に対しいくらかの進歩をもたらし，原子や元素といった基本的な考えを導いた。化学に先んじた錬金術は中世に栄え，化学の知識に寄与したが，その秘密性の故に，知識は効果的に伝播せず，進歩は遅くなった。

3. 16世紀になって，科学者は自然界を理解するためのいとぐちとして観察に焦点をあわせた。コペルニクスとベサリウスが著した本は，この考え方の変化を例証しているし，科学革命の始まりを印している。これらの本は，化学の世界についての我々の理解の比較的急速な発展に引き継がれた。ボイルは物質を組成による分類に使用できる式を提案した。ラボアジェとプルーストは，それぞれ，質量の保存と定比例組成の法則を公式化した。ジョン・ドルトンはこれらの法則に基づいて，原子論を樹立した。その後ラザフォードは原子の内部構造を調べ，原子はほとんどが空の空間であり，中心に非常に密度の高い核，そして，その周りを回る負電荷の電子から成ることを見出した。現代化学の基礎が置かれたのである。

4. 化学を理解することは，我々の世界の理解と我々自身の理解を深める。なぜなら，すべての物質は，我々の脳や身体さえも，原子と分子でできているからである。

5. ギリシャ人の考え方は社会に影響を与えた，いやむしろ，科学がもたらすであろう影響を遅くした。ほとんどの人がこの方法を疑わなかったが，その結果として，科学と技術は数百年間大きく無視された。これを疑わしく思うならば，もしギリシャ人の間で紀元後600年に科学革命が起こっていたとすると，我々の世界はどのようになったであろうか，考えてごらんなさい。我々の社会は，科学の進歩が始まってからたった450年の後で顕著に変わってきた。2600年後には，社会はどこにいることになるだろうか。

6. 観点のシフトから生じた社会の変化は大きかった。我々が物理的世界をいかに理解し制御するかを知り始めるとともに，科学は栄え始めた。科学的方法を通して獲得した知識を人の生活の改良に応用し始めるにつれ，技術も成長した。これを疑うなら，自分自身の生活を考えてごらんなさい—技術なしでの生活はいかに違っているだろうか。これらの変化とともに，知識と技術を賢く使う責任が生じた—全体としての社会だけがこれができる。科学が我々に授けた力は社会を良くするために使われてきた—例として，医学の進歩を考えて見なさい—しかし，それはまた破壊するのにも使われてきた—原子爆弾や公害を考えてごらんなさい。我々は，社会として，科学が善のために与えた力をどのように利用し，どのように悪を避けるか。

[Ex. 26] ＳＩ系

　世界中ほとんどの人々は国際単位系と呼ばれる一連の測定単位を用いる。これは通常 SI 系と引用される。SI 系は，古いメートル系を新たに修正したものである。SI 系が広く使われるのは，10 進法であるためである。あらゆる単位は，次に小さい単位の 10 倍である。例えば，1 センチメートルは 10 ミリメートルである。使い慣れた英国の測定系は 10 進法ではなく――1 ガロンは 4 クォート，1 フィートは 12 インチ，1 ポンドは 16 オンスである。

　SI 系は，少数の基本単位(例えばメートル)と一連の接頭語(例えば，キロ，センチ，ミリ)を用い，接頭語はすべての単位につけて用いることができる。化学者は，古いメートル単位系，例えばリットルやミリリットルのように，SI 系ではない単位を続けて利用している。体積 1 リットルは 1 立方デシメートル($1\,dm^3$)であり，1 ミリリットルは 1 立方センチメートル($1\,cm^3$)である。

　測定量は，常に 2 つの部分，すなわち数字と単位から成っている。諸君は，自分の測定値に単位を書く習慣をつけなければならない。ある長さを，5 ではなく，5 mm，5 cm，5 m などと書かなければいけない。

[Ex. 27] 周 期 律

　1. 1980 年代，ドミトリ・メンデレーエフ(1834-1907)という有名なロシアのペテルブルグ工科大学の教授が化学のテキストを書いた。記述化学の育ちつつある知識を描いて，元素が同じような性質をもつ種族に分類できることに気づいた。ヘリウム，ネオン，アルゴンのようないくつかの元素はすべて化学的に不活性な気体である；すなわちこれらは反応して化合物を作らなかった。他のものは，ナトリウムやカリウムのように，反応性の金属であった。

　2. さらに，彼は，原子量の増加する順に元素を表にすると，これらの同じような性質が周期的な方法で繰り返すことを見つけた。メンデレーエフはこれらの観察を周期律にまとめた。これは，「元素が原子量の増加する順に並べるとき，ある性質の組が周期的に繰り返される」と述べている。

　3. メンデレーエフは，そこで，すべての知られている元素を表に整理し，原子量が左から右に増加し，同じ性質を持つ元素が同じ縦の列に並ぶようにした。これがうまくいくために，メンデレーエフは彼の表に空所を残さなければならなかった。彼は，これらの空所を埋めるような元素が発見されるだろうと予測した。彼はまた，いくつかの測定された原子量が誤りであることを提案しなければならなかった。両方の場合とも，メンデレーエフは正しかった。

　4. メンデレーエフの提案の 20 年以内に，3 つの空所がガリウム(Ga)，スカンジウム(Sc)，ゲルマニウム(Ge)の発見で埋められた。メンデレーエフの元素の配列は，周期表と呼ばれ，現代化学の基本となっている。

　5. メンデレーエフは，周期律がなぜ存在するのか分からなかった。彼の法則は，すべての科学の法則と同じように，数多くの観測したことをまとめていたが，観測された振る舞いの基となる理由は与えていなかった。科学的な方法の次の段階は，この法則を説明し，原子のモデルを与えるような理論を案出することであった。メンデレーエフの時代には，周期律を説明する理論は何も無かった。

　6. なぜ元素の性質が周期的に繰り返されるのかを説明する理論を垣間見てみよう。その理論は，ニールス・ボーア(1885-1962)によるので，原子のボーアモデルと呼ばれる。これは，あるいくつかの元素が同じような性質を繰り返して示すという巨視的な観察を，元素を形成する原子

が繰り返す類似性を持っているという微視的な理由に結び付けている。

[Ex. 28] 放 射 能

1. ベクレルとキュリー夫妻によって発見された放射能は，不安定な核から発せられた高エネルギーの粒子から成る。アルファ線は，ヘリウム核から成り，イオン化力が高く，透過力は低い。ベータ線は，原子核内の中性子が陽子に変換する時に発せられる電子から成る。ベータ粒子はアルファ粒子よりもイオン化力が低いが，透過力は高い。ガンマ線は，イオン化力が低いが，透過力が高い高エネルギー電磁線である。不安定原子核は，その半減期，すなわち与えられた試料中の核の半分が崩壊するのにかかる時間，に従って放射性崩壊する。

2. U（ウラン）-235やPu（プルトニウム）-239のような，いくつかの重元素は，中性子で衝撃されると不安定になって，核分裂を行う。原子は分裂して，軽元素，中性子，およびエネルギーを生じる。核分裂を制御下に保つと，生じたエネルギーは発電に利用することができる。もし，核分裂をエスカレートさせると，原子爆弾になる。水素爆弾は，太陽と同じで，核融合と呼ばれる別の形の核反応を使うものであり，そこでは軽元素の核が結合して重元素になる。エネルギーを生じるすべての核反応で，いくらかの質量がこの反応中にエネルギーに変換される。

3. 化石や岩石中のある放射性元素の量を測定することによって，放射能は物体の年代を推定するのに利用できる。地球の年代は，最古の岩石中のウランと鉛の比に基づいて，45億年と推定される。高レベルの放射能は人命を殺し得る。低レベルは，病気の診断とか処置といった治療の形で利用される。

4. 放射能の発見は，私たちの社会に多くの影響をもたらした。それは，ついにはマンハッタン計画，すなわち1945年の最初の原子爆弾の建設と爆発にまで導いた。社会は，初めて，非常に明確な方法で，科学がもたらした力の威力を見ることができたのである。しかし，科学自体が日本に爆弾を落としたのではない。それをしたのは合衆国の人々であり，次の問題が残っている：技術が与える力を私たちはいかに利用するか？　それ以来，私たちの社会はある科学的発見の倫理的な意味と苦闘してきた。この10年間，核兵器は年に2,000個の爆弾の割合で軍備縮小されてきた。今日，私たちは，核絶滅の恐れがそれほど深刻ではない時代に住んでいる。

5. 核分裂は，化石燃料の燃焼に関連した危険な副作用もなく，発電に利用される。それでも，原子力はそれ自体の問題，すなわち，事故や廃棄物処理の可能性を抱えている。合衆国は核廃棄物処理の永久的な場所を建設するだろうか。化石燃料の供給が減少するにつれて，我々は原子力に変えることができるだろうか。どれだけの資源を，未来のエネルギー源として核融合の発展に置き換えるだろうか。これらは，すべて，この新しい千年（2001～3000）を始めるときに私たちの社会が直面する疑問である。

6. 核過程は，私たちがどれほどの年であるか（人がどれほど長い歴史をもっているか）を語ることができた。考古学的発見は，年代順の謎に合わせて，最古の時代からの人類の歴史について物語っている。私たちは，人類が生存するよりも前に地球上で何十億年も過ぎたことを知っている。私たちは，人がどのように道具を使い始め，地球上をどのように移住し，移動したかを知っている。私たちは，トリノの経かたびらのような特定の品目の年代を推定し，それらが本物かどうか決めることができる。科学的な観点は，私たちの社会にどんな効果を持つだろうか。宗教にはどうであろうか。それは，私たちが何者であるかについて，何を語るのであろうか。

[Ex. 29] 炭素の同位体

1. 天然の炭素は，3つの同位体を含む：最も多い同位体であるC-12（98.8%）；少量のC-13（1.11%）；および，微量のC-14。C-14は半減期5,700年を持つ放射性同位体である。このような短い半減期であると，我々はこの同位元素が地球上にほとんどそのきざしがないことを予想する。しかし，それはすべての生物組織に行き渡っている，というのは，この同位体は高層大気中で宇宙線の中性子と窒素原子の間の反応で常に作られているからである：

$$^{14}_{7}N + ^{1}_{0}n \longrightarrow ^{14}_{6}C + ^{1}_{1}H$$

2. 炭素原子は酸素と反応して，二酸化炭素の放射性分子を生成する。これらは光合成で植物に吸収される。植物を食べる動物と，植物を食べた動物を食べる動物は，すべて，同じ割合の放射性炭素を含み，すでに体中にある炭素-14は崩壊する。このように，ある対象物の年代は，試料中に存在する炭素-14を測定することで決定できる。この方法は，1,000年から20,000年の間の対象物の年代を測る絶対尺度を提供する。W. F. Libbyは，この放射性炭素年代測定技術を開発したことで，1940年ノーベル化学賞を受賞した。

[Ex. 30] 二酸化炭素，ヘンリー則，絶対零度

1. 二酸化炭素ガスは食物消化の最終生成物の1つであり，我々が吐き出す空気の中にあるが，吐き出された息は，たった4%のCO_2である。我々が呼吸する大気は0.03%CO_2を含む。血液中の二酸化炭素の存在は呼吸を刺激する。この理由で，人工呼吸の酸素や麻酔に用いられるガスに，二酸化炭素が加えられる。

2. 二酸化炭素は，また，どんな有機（炭素を含む）燃料が燃えるときにも生成される。ガソリン，石炭，天然ガスのような燃料，木や紙，すべてが，燃えるときにCO_2を生じる。あまりに多くのこれら燃料を燃やすと，我々の大気にあまりに多くの二酸化炭素を加えることになるだろう—これは，地球温暖化として知られている過程で地球の過熱に導き得る。植物は，空気から二酸化炭素を取り除き，それから光合成過程で食物を作る。二酸化炭素はまた，海水中に溶けることによって大気から取り除かれる。

3. ソーダのような炭酸飲料は，溶けた二酸化炭素を含んでいる。クッキーやケーキを焼く際に，化学反応が重曹や膨らし粉からCO_2を生成する。二酸化炭素ガスは，クッキーやケーキ中に細かな穴を作る。パンやピザ生地を焼くときに，酵母が砂糖から二酸化炭素を生成する—そのCO_2が生地をふくらませる。

4. 二酸化炭素ガスは火を消すことができる。それがこのようにできるのは，燃えている物体の周りの空気中の酸素量を減らすことによる。空気中の酸素は火に不可決であり，酸素が取り除かれると，火は消える。

5. 二酸化炭素は室温で気体であるけれども，−78.5℃以下の温度では固体である。固体の二酸化炭素は「ドライアイス」と呼ばれる。それは固体から気体へ，決して液体を生成することなく変わるからである。すなわち昇華と呼ばれる過程である。

6. [ヘンリー則] ヘンリー則によれば，気体の液体中への溶解度は，液体上の気体の圧力に正比例する。圧力を増すと，気体の溶解性を増し，圧力を減らすと，この溶解性を減じる。炭酸飲料の開けていないボトルは，その溶液中に溶けた大量の二酸化炭素ガスを含む。溶けているときは，ガスは見えない—つまり，混合物は均一である。炭酸飲料の開けたボトルは，二酸化炭素

の泡が表面に上がってくる。これらの泡は溶けていない。ボトルを開けることは圧力を下げ，水中への二酸化炭素の溶解度を減らすのである。

7. ［絶対零度］ 温度は，物質試料中の分子の平均の運動エネルギーの尺度である。室温では，分子は非常に早く動いている。分子が遅くなると，その温度は減少する。理論的には，ある非常に低い温度で，分子の運動は遅くなってゼロになり，分子は完全に運動のないものになる。絶対零度といわれる温度は，-273.15℃すなわち0Kに等しい。科学者は最近絶対零度に非常に近い0.000000002Kに到達した。最近の理論は，絶対零度は達することはできないと言う；分子運動は完全に止むことはできない。

8. 硬い容器（容積が一定のままであるもの）の中では，温度の増加は気体の圧力を増加させる。温度の低下は圧力を下げる。この関係はしばしば圧力―温度の法則として知られている。これは，容積が一定であるという条件で，気体の圧力は絶対温度とともに変化することを述べている。実験では，圧力計を付けた中空のスチール球に空気を満たす。この球をさまざまな温度の液体中に浸して，球内の圧力を測定する。グラフ用紙に温度と圧力をプロットして，最適の直線を引く。この線を外挿する。絶対零度は，圧力がゼロになる温度である。

[Ex. 31] イオン性および共有結合性化合物

1. 自然にあるほとんどの物質は，化合物，すなわち決まった比率での元素の組合せである。化合物は化学式で表され，これは少なくとも，存在する各元素の結合様式と相対量を同定するものである。共有結合化合物に対しては，分子式と呼ばれる，もっと明確な種類の化学式が各分子中の原子の数と結合様式を示す。

2. 化合物は2つの結合様式すなわちイオン性と共有結合性に分類され，それぞれは独自の命名法をもっている。イオン性化合物は，イオン結合を介して非金属に金属が結合している。イオン結合では，電子が金属から非金属へ移動されて，金属をカチオン（正に荷電），非金属をアニオン（負に荷電）にする。固体の形では，イオン性化合物は，正と負イオンが交互になった三次元格子から成る。共有結合化合物は，共有結合を介した非金属と非金属の結合である。共有結合では，電子が2つの原子間に共有される。共有結合化合物は，分子と呼ばれ，同定できる原子のかたまりに等しい。分子の性質は，それらが構成する共有結合化合物の性質を決める。

3. 分子式中のすべての原子の原子量の総和は，分子量と呼ばれる。それは，化合物の質量とその分子のモル数の間の変換因子である。化学反応では化合物が生成したり，変化したりするが，それは化学式で表される。化学式の左辺にある物質は反応物と呼ばれ，右辺にある物質は生成物と呼ばれる。化学式の各辺にある，各タイプの原子の数は，式が釣り合うように，等しくなければならない。化学式の係数は，反応物と生成物の量の数的関係を決める助けとなる。

4. 私たちの周りにあるほとんどの物質は化合物であるので，私たちの周りで何が起こっているかを理解するために，それら（化合物）を理解しなければならない。化合物，あるいは化合物を構成する分子は，私たちが毎日使う物質，例えば，プラスチック，洗剤，あるいは発汗抑制剤から，私たちが社会として直面する環境問題，例えば，オゾンの枯渇，空気汚染，あるいは地球温暖化に至るまで，すべてにおいて重要である。

5. イオン性化合物は，食品―例えば，食塩はNaClである―および海水や土の中で見出される。共有結合化合物は，水，私たちが燃やす燃料，食べる食品の多く，そして生活に重要な分子のほとんどを構成する。

6. 化学反応によって私たちの社会も私たち自身の身体もともに（活動を）続けていると言って

も控えめな表現ではない。私たちのエネルギーの 90 パーセントは化学反応，主として化石燃料の燃焼から由来する。私たち自身の身体も，私たちの食べる食品からエネルギーを引き出す。それは食品に含まれる分子のゆっくりとした燃焼を編成することによる。

7. 有用な化学反応の生成物は，ときどき環境問題を起こすこともある。例えば，化石燃料燃焼の生成物の 1 つである二酸化炭素は，温室効果と呼ばれる過程を通じてこの惑星が温暖化するのを引き起こしている可能性がある。

[Ex. 32]　エネルギーと仕事

力が物体に作用し，その物体を動かすとき，物体の運動エネルギーの変化は，その物体になされた仕事に等しい。例えば，車を時速 0 から 60 マイルに加速したり，野球のボールを場外に打ち出すのには，仕事がなされねばならない。物体の位置エネルギーを増すためにも仕事が必要である。例えば，物体を重力に逆らって上げたり（エレベーターのように），ナトリウムイオン（Na^+）を塩素イオン（CL^-）から離したり，電子を原子核から取り去るために，仕事が必要とされる。物体になされた仕事は，その物体に移動されたエネルギー量に相当する。すなわち，仕事をすることは物体にエネルギーを移動する過程である。逆に，もし物体が他の何かに仕事をするならば，その物体に結びついたエネルギーは減少しなければならない。

[Ex. 33]　化学熱力学への導入

1. 利発的な若い科学者は，自分の専門分野に初歩的なバックグラウンドをもったばかりであっても，「熱力学」という科目が化学，生物学，物質科学，地質学と何らかの関係があると分かって，驚くかもしれない。熱力学という言葉は，文字通りにとれば，熱によって生じる機械的な作用に関係した分野を意味する。

2. Kelvin（ケルビン）卿（William Thomson 本名：ウイリアム・トムソン）がこの言葉を考案したのは，熱の動的な性質に注意を向け，この見方を，熱を流体の 1 つと見る従来の概念と対比させるためであった。Kelvin がこの名前を創ったときよりも，この科学の応用ははるかに広くなっているけれども，この名前はそのまま残っている。

3. 力学，電磁場理論，や相対性理論，そこでは Newton（ニュートン），Maxwell（マックスウェル），Einstein（アインシュタイン）という名前が突出しているのと対照的に，熱力学の基礎は，6 人を超える個人の思索から始まっている：Carnot（カルノー），Mayer（マイヤー），Joule（ジュール），Helmholtz（ヘルムホルツ），Kelvin（ケルビン），Clausius（クラウジウス）などである。それぞれの人達が重大な段階を提供して，2 つの古典的な熱力学法則という壮大な統合体に導いたのである。

[Ex. 34]　化学熱力学の目的

1. 実際上，化学熱力学の主な目的は，与えられた物理的あるいは化学的変化の可能性あるいは自発性を決める基準を確立することである。例えば，私たちは，1 つの相から別の相への自発的変化，例えば，グラファイトからダイヤモンドへの変換とか，細胞中で起こる代謝反応の自発的な方向など，の可能性を決める基準に興味がある。ギブスのエネルギー関数で表される，熱力学の第 1 および第 2 法則を基に，いくつかの理論的概念と数学的関数が開発されて，これらの問題解決に強力な方法を提供している。

2. ある自然過程の自発的方向がいったん決まると，私たちは，その過程が平衡に到達する前

にどこまで進行するかを知りたくなる．例えば，ある産業工程の最大収率や大気中の二酸化炭素の天然水中への平衡溶解度，あるいは細胞中における一群の代謝物質の平衡濃度などを知りたくなるであろう．熱力学の方法は，このような量を推定するのに必要な数学的関係を提供している．

3. 化学熱力学の主な目的は自発性および平衡の解析であるけれども，その方法は多くの他の問題にも応用できる．例えば，理想系および非理想系における相平衡の研究は，抽出，蒸留，結晶化の聡明な利用や，冶金操作，新材料の開発，地質系で発見された鉱物種の理解に，基本的なものになっている．同様に，物理的あるいは化学的変化に伴うエネルギー変化は，熱または仕事のどちらの形であれ，非常に興味深いことである．この変化が燃料の燃焼でも，ウラン核の分裂でも，あるいは濃度勾配に逆らった代謝物質の輸送であっても，熱力学の概念と方法は，このような問題を理解する強力な手がかりを提供する．

[Ex. 35] 熱力学の第1, 第2, 第3法則

1. クラウジウスは熱力学の発見を，「世界のエネルギーは不変である；世界のエントロピーは最大値に向かう」という一文に要約した．そして，ギブスの偉大な回想録「不均一物質の平衡」の冒頭になったのがこの引用文であった．このような大家たちがエネルギーと同格の重要な位置においたが，熱力学を学ぶ非常に多くの学生に怖いものとなった，このエントロピーとは何だろうか．

2. 熱力学の第1法則，すなわち，エネルギー保存の法則は，それが記述されたとほとんど同時に広く受け入れられたが，その理由は，それを支持する実験的証拠がその当時圧倒的であったというよりは，むしろ，それが合理的であり人の直感に一致しているように思われたからであった．物事の永続性に関する概念は，すべてに共有されているものである．それは物質の世界から精神の世界に至るまで拡張されている．物体が破壊されたとしても，その物質は何らかの方法で保存されるという考えは，古代人によって私たちに伝えられてきた．そして，現代の科学で，このような考え方の有用性は完全に理解されている．炭素の保存の認識は私たちに，少なくとも思考の中で，石炭が燃やされたとき，生じた二酸化炭素が生きた植物に吸収され，そこからこの炭素は終りのない一連の複雑な変化を経るという，この元素の過程を辿らせる．

3. 熱力学の第2法則は，エネルギーの散逸あるいは退化の法則，あるいはエントロピー増加の法則としても知られているが，カルノー，クラウジウス，そしてケルビンの基礎的研究を通じて，第1法則とほとんど同時に開発された．しかし，それは違った運命をたどった，というのは，それは現存(当時)の思考や先入観と一致しているようにはどうしても見えなかったからである．さまざまな保存則は科学的思想の本体に受け入れられるよりもずっと前から予示されていた．第2法則は，一般的な宇宙論に関する広範な内容を含んでいる新しいこととして，伝統的な思想には相容れないものものとして登場した．

4. 第2法則は直感と異なるように見えたし，さらに，当時の哲学には相容れないものとさえ見えたために，この法則への例外を見つけて，その普遍的正当性を反証する試みがなされた．しかし，このような試みは，むしろ，懐疑派を納得させ，熱力学第2法則を近代科学の基盤の一つとして確立するのに役立った．この過程で，我々はその哲学的含蓄にまで理解がおよぶようになり，それらを満足いくまで解釈することを学んだ；我々はその限界を学んだ，あるいは，もっと良く言えば，我々は，これらの限界は最早存在するように見えないという形で，この法則を述べることを学んだ．そして，特に，我々は，他の良く知られている概念との関係を学び，（その

結果）それは今やかけ離れたものとしてあるのではなくて，むしろ長く親しんできた考えの当然の帰結であったのである。

5. 19世紀の間に，熱力学の第1および第2法則が定式化され，多くの科学的問題に応用された。（しかし）20世紀の最初の10年までは，極低温での研究が十分に進歩しても，それが熱力学の第3法則を構成している概念の基礎へ発展するまでには至らなかった。それでも，今では，第3法則は数多くの厳しい実験的試験にも満足な答を出し，化学反応の大きな部分のデータの基礎を提供している。したがって，第3法則を，我々の主題の基本的土台に欠くことのできない部分として含めるのが，適当と思われる。

6. 我々は，固体の熱容量は，絶対零度に近づくにつれて，急速に減少することを見てきた。また我々は，固体の内部エネルギーは量子化されていて，有限のエネルギー量子だけが吸収され得るというアインシュタインの説明を思い出す。それゆえ，固体が0Kに冷却されると，すべての成分粒子はその最低の量子状態に落ちる；実際，完全な結晶がその最低量子状態にあると言う事ができる。これは明らかに特に単純な状態で，これを確率の絶対単位（訳注：確率1，エントロピー0）ととる事ができる。

[Ex. 36] 成分の数と選択

1. ギブスは書いた：「条件が満たされれば，考慮中の均一な集団の成分と見なすべき物質の選択は，全く都合の良いように，集団の内部組成に関するどんな理論にも無関係に選ばれる」と。条件とは，成分の数pを含み，成分の微分$dn_1, dn_2, \cdots, dn_p$の値が「独立で，考慮している均一集団の組成のあらゆる可能な変化を表している」ようなものでなければならない。この原理にしたがって，考慮中の相の中に存在することが分かっている物質，または存在することが想像される物質のポテンシャルμ_iを定義することが可能である。ただし，これらμ_iの値のいくつかは同じであることが判明するかもしれないし，いくつかはいろんな形で関係しているかもしれない。

2. 成分の数に関しては，観測のタイムスケール（時間尺）が重要である。室温で，触媒のないときは，水素，酸素，水の混合物は，人の観測可能な時間内で何の相互変換もないので，3成分系である。高温あるいは触媒存在下では，2成分系である。なぜなら，相互変換が十分速く，人の観測可能な時間で，水の割合は系の最終組成で決まるからである。中間の条件では，系の挙動は，速い観測では3成分系に近づき，遅い時間尺では2成分系に近づく。

[Ex. 37] 管流動反応器

1. 反応物の混合と時間間隔の測定という，対になる問題は，半減期が1時間以上の反応，すなわち，最初に存在する反応物の半分の反応に必要な時間が1時間もあるような反応では，その研究に重大な困難は何もない。半減期が1時間から1秒の間にある反応に対しては，比速度の精確な評価の方法に困難（しかし，克服できないものではないが）を伴う。速い反応では，溶液の組成に素早く適応するような物理的性質を時間の関数として記録するような自動装置を使って反応率を評価することにより，時間の問題は軽減することができる。この性質の値が反応率と直線関係にあるなら，かなりの利点である。

2. この種の測定の長所を利用し，同時に混合の問題を軽減する1つの有効な方法は，管流動反応器である。HartridgeとRoughtonによるこの装置の開発は，上手く測定できる反応速度の制限を0.001秒ほども小さい半減期の系にまで拡げ，以前にはまったく近づけなかった反応速度論の領域を開いた。操作の技術は高度に開発され，多くが自動化の方法でなされ，その結果は非

常に重要になってきた.

3. この技術では，一定の速度で流動する2つの溶液が急速にかつ完全に混合されて，直径が一定の管の中を通過させる．好条件のもとでは，混合点からの距離 l における薄い管状の溶液薄層の組成は，流動の無い条件で，混合液が $t = Al/u$ で与えられる時間だけ反応したとしたときと同じである：ここで，A は管の断面積，u は流速(単位時間当たりの体積)である．したがって，反応率に依存する物理的性質を距離 l の関数として測定すると，非流動系でその性質を時間の関数として測定したのと同じ反応動力学についての情報が提供される．

4. この種の測定で決定的に重要なことは，溶液の組成が管に直交する全ての断面を通じて実質的に一定であること，管軸に平行方向の混合と拡散が無視できること，そして特に，入れた溶液の混合が実質的に完全となった時点で観測を始めるべきことである．混合が不完全な系では，ごた混ぜ状態の，さまざまな組成の体積要素内で反応が起こり，観測された反応率は無意味となる．

[Ex. 38] 両親媒性化合物としての石けん

表面活性剤は，ジキル博士とハイド氏のように，相反する性質を持った2つの部分から成る．1つの部分は親水性であり，他方は疎水性である．典型的な表面活性分子，オクタデカン酸ナトリウム(石けんの主成分の1つ)は下に示されている．図で分かるように，表面活性分子は，通常マッチのように描かれる，マッチの頭が親水性基(石けんの場合は，カルボン酸ナトリウム基)，軸が疎水基である．疎水基は通常炭化水素鎖であり，したがって，しばしば親油基と呼ばれる．表面活性を示す物質はいくつかの名前で呼ばれる—表面活性物質，表面活性化合物，そして表面活性剤(サーファクタント)のように．表面活性物質は水と油の両方に親和性を示すので，それらはまた，両親媒性化合物，両親媒性種，あるいは，両性化合物などとも呼ばれる．

[Ex. 39] 表面張力の源

凝縮した物質(液体と固体)は，それらの分子間に凝集エネルギーが存在するために，表面張力をもっている．凝縮物質の内部の分子はその周りの分子と引力で相互作用する．例えば，内部液相中の水分子は，ファンデルワールス相互作用と同様に，いくつかの(最高4つの)水素結合を持ち，ダイヤモンド結晶中の炭素原子は最近接の原子と4つのC-C共有結合を持っている．しかし，表面にある分子は，このような結合および／または相互作用を完全には作ることができない．なぜなら，それらは真空(蒸気)側には相互作用する分子を持たず(あるいは，ほんの少ししか持たず)，したがって，内部相にある分子と比べて，過剰のエネルギーを持つ．この表面分子あるいは原子に存在する過剰エネルギーが表面張力として定義される．

[Ex. 40] 電磁放射

1. 有機化合物の特性化技術のほとんどに共通するのは，電磁放射である．電磁放射というのは光である．我々の見る光だけでなく，見ることのできない光でもある．電磁スペクトルは連続的なエネルギーの範囲を表すけれども，それは，エネルギーの減少する順あるいは波長の増加する順に，γ-線，X-線，紫外，可視，赤外，マイクロ波，そしてラジオ波に分けられる．

2. 電磁照射は，周波数，波長，振幅によって特性化される．光のエネルギーは，その周波数 ν，すなわち，単位時間当たりにある固定点を通過する波の数，に比例する．沢山の波は沢山のエネルギーを意味する．したがって，より短い波長の光は，エネルギーがより高い，というの

は，より沢山の波が単位時間に，ある固定点を通過し得るからである。周波数はヘルツ（Hz）で計られ，「秒当たり」すなわち s^{-1} に対応する。例えば，多くの携帯電話の操作周波数である1秒当たり 900,000,000 波，すなわち 900 MHz などである。

3. 物質が原子と呼ばれる不連続な単位のみでできているのとちょうど同じように，電磁エネルギーも量子と呼ばれる不連続な量でのみ伝達される。エネルギー量 ε は，与えられた周波数 ν の1量子のエネルギー（1光子）に相当し，プランク式で表される：$\varepsilon = h\nu$：ここで，

$h =$ プランク定数 $(6.62 \times 10^{-34}$ J・s $= 1.58 \times 10^{-34}$ cal・s$)$。

4. 1つの波の長さ，すなわち波長は，すべての波が同じ速さで移動するので，計算することができる。波は，電磁放射の速さ，すなわちもっと普通に言えば，光の速度で動く。光速を c で略す。これは 3×10^8 m/s に等しい定数である。波長 λ は，周波数あるいはエネルギーから次式で計算できる：$\lambda = c/\nu$　または $\lambda = hc/\varepsilon$

5. 我々は電磁スペクトルのさまざまな部分を利用して，分子の構造についてさまざまなタイプの情報を得る。

X線結晶学。高エネルギー，短波長のX線が分子の周期的格子あるいは結晶を通過させられる。光は結晶中の原子の面から回折し，この散乱から三次元構造が計算される。

6. 紫外（UV）および可視分光学。広いスペクトル（UV-可視範囲）の光が試料分子中を通過させられる。π結合の共役ネットワークはこの中程度のエネルギーの光を吸収する。共役π結合ネットワークの違いはスペクトルの違う領域の吸収を導くので，これから構造が同定される。しばしば，1個の幅広い特徴的な吸収が診断用（同定用）となる。

7. 赤外（IR）分光学。これは，より低エネルギー範囲の電磁スペクトルが用いられることを除けば，UV-可視分光学と同様に作用する。試料に吸収されたスペクトル部分は，分子中の特定の結合の振動に対応する。UV-可視スペクトルと違って，ある分子のIRスペクトルは多数の吸収を含み，これらは分子中のあらゆる可能な振動に対応している。この情報の少しの量だけが特定の官能基を同定するのに必要とされる。

8. 核磁気共鳴（NMR）分光学。炭素と水素を含む有機分子が強い磁場中におかれるとき，ラジオ波を利用して，これらの原子の接続性を探ることができる。核磁気共鳴（NMR）スペクトルから，炭素と水素の配置が決定できる。

[Ex. 41]　紫外（UV）および可視分光学

1. 紫外（UV）は，10^{-8} m から可視領域の低波長末端（4×10^{-7} m）まで拡がっている。しかし，有機化学者の最も興味のある部分は，2×10^{-7} m から 4×10^{-7} m までの狭い範囲である。この領域の吸収は，ナノメートル（nm）で計られる，ただし，1 nm $= 10^{-9}$ m $= 10^{-7}$ cm である。したがって，興味のある紫外領域は 200 から 400 nm である。UV 照射のエネルギーは不飽和分子のπ電子のエネルギー準位を上げるのに必要な量に相当する。

2. UV スペクトルは，試料に連続的に変化する波長の UV 光を照射することで記録される。波長が不飽和分子中のπ電子を高い準位に昇位させるのに必要なエネルギー量に対応すると，エネルギーが吸収される。この吸収が検知され，波長と吸収された放射エネルギー量をプロットしたチャート上に表示される。

3. 一般に多くのピークを持つ IR スペクトルと違って，UV スペクトルは通常極めてシンプルである。しばしば，単一の幅広いピークだけがあって，これがその頂点の波長（λ_{max}）を記録して同定される。例えば，ブタ -1,3- ジエンでは，$\lambda_{max} = 217$ nm である。

4. 不飽和分子の π 電子のエネルギーを上げるのに必要な照射の波長は，分子中の π 電子の性質に依存する。最も重要な因子の一つは共役の程度である。電子遷移に必要なエネルギーは，共役の程度が増すとともに減少することが判明している。したがって，上記のブタ-1,3-ジエンと比較して，ヘキサ-1,3,5-トリエンは λ_{max} = 258 nm で吸収し，オクタ-1,3,5,7-テトラエンは λ_{max} = 290 nm を持つ。

5. ジエンやポリエンに加えて，ほかの種類の共役 π -電子系も UV 吸収を示す。ブテノン(but-3-en-2-one, λ_{max} = 219 nm)のような共役エノン，およびベンゼン (λ_{max} = 254 nm)のような芳香族分子も，構造決定の助けとなる特徴的な UV 吸収を持つ。

6. UV-可視分光学は，溶液中の種(物質)の濃度を測定する普通の技術である。Beer 則は次のように言う：試料の吸光度 A は，溶液中のその分子の濃度に比例する： $A = \varepsilon c l$ ：ここで，ε = 分子のモル吸光係数で，分子が光をどれだけ吸収するかを反映する分子特有の定数，単位は $L\ mol^{-1}\ cm^{-1}$ (通常，その化合物 λ_{max} に対して与えられる)，である，c = 溶液中の分子の濃度 (mol/L) そして，l = 試料中の光の通る距離，すなわち光路長 (cm) である。

[Ex. 42]　赤外分光学

1. IR 領域は，可視の丁度上(波長 7.8×10^{-7} m)から，およそ 10^{-4} m までの範囲をカバーするが，中間の領域(2.5×10^{-4} から 2.5×10^{-3} cm)だけが有機化学者によって利用される。波長の代わりに波数が通常用いられる。これは，波長の逆数(cm^{-1})で与えられる。したがって，通常の IR スペクトルは 4,000 から 400 cm^{-1} に及ぶ。

2. 分子は，なぜ，ある波長の IR エネルギーを吸収して，他は吸収しないのか？すべての分子は，ある量のエネルギーを持っていて，これが結合を伸ばしたり，縮めたり，原子を前後に振らしたり，その他の分子運動を引き起こす。分子が含むエネルギー量は連続的に変わり得るのではなくて，量子化されている。すなわち，分子は，特定のエネルギーレベルに対応した特定の振動数だけで振動するのである。

3. 例えば，結合の伸縮をとってみよう。我々は，結合長をあたかも固定されているかのように話すけれども，与えられた数字は現実には平均である。というのは，結合は常に伸びたり曲がったり，長くなったり縮んだりしているからである。したがって，典型的な C—H 結合は平均結合長 110 pm を持つが，実際にはある特定の振動数で振動し，あたかも 2 つの原子を結ぶばねがあるかのように，交互に伸びたり，縮んだりしているのである。

4. 分子が電磁放射で照射されるとき，放射線の周波数が振動の周波数と一致すると，エネルギーが吸収される。エネルギー吸収の結果は，その振動の振幅の増加である；言い換えると，2 つの原子を結ぶ「ばね」がさらに幾分伸び縮むのである。分子によって吸収されるそれぞれのエネルギーはある特定の分子運動に相当するので，我々は IR スペクトルを測定することによって，分子がどのような種類の運動を持っているかを知ることができる。それでこれらの運動を解釈することによって，どのような種類の結合(官能基)が分子中に存在するかを見出すことができる。

5. IR スペクトルの完全な解釈は，大抵の有機分子は大きくて，多数〈数ダース〉のさまざまな伸縮と変角の運動があるので，難しい。したがって，IR スペクトルは通常多数の吸収を含む。幸い，有用な情報を得るのに IR スペクトルを完全に解釈する必要はない，というのは，官能基は特性 IR 吸収を持ち，これらは化合物によって変わることはない。ケトンの C=O 吸収はほとんど常に 1,680 から 1,750 cm^{-1} の範囲にあり，アルコールの O-H 吸収はほとんど常に 3,400 か

ら 3,650 cm^{-1} の範囲にある，などである．どこに特徴的な官能基吸収があるかを学ぶことで，IR スペクトルから構造情報を得ることが可能である．

[Ex. 43]　核磁気共鳴画像解析診断（MRI）

1. 有機化学者が常に行っているように，NMR 分光学は構造決定の有力な方法である．少量の試料，典型的には数ミリグラムかそれより少ない量を凡そ 1 mL の適当な溶媒に溶かして，溶液を細いガラス管に入れ，このチューブを強磁場の両極間の細い（1-2 cm）隙間に置く．しかし，もっとはるかに大きい NMR 機器があったと想像してみよう．数ミリグラムの代わりに，試料量は数 10 キログラム，磁場極間の細い隙間の代わりに，その空間は十分大きくて，人が乗り込んで，身体部分の NMR スペクトルを得ることができる．今想像したものが磁気共鳴画像診断（MRI）の機器である，これは，X 線や放射線画像診断法に勝る利点のために，医学における非常に重要な診断技術である．

2. NMR 分光学と同様に，MRI はある種の核，特に水素の磁気的性質と，これらの核がラジオ周波数エネルギーによって刺激されたときに発生するシグナルを利用する．ただ，NMR 分光学で起こることと違って，MRI 機器は強力なコンピューターとデータ処理技術を用いて，核の化学的性質よりは身体中の磁気核の三次元的位置を見る．大抵の MRI 機器は現在水素を観測する．水素は，身体中の水や脂肪のあるところには豊富に存在するからである．

3. 生じるシグナルは，水素原子の密度とその周りの性質とともに変わるので，さまざまなタイプの組織の同定を可能にし，さらには，動きの可視化をも可能にする．例えば，1 回の鼓動で心臓を出る血液の体積が測定できるし，心臓の動きが観測できる．X 線ではよく現れない軟らかい組織もはっきりと観測され，脳腫瘍，卒中や他の状態の診断を可能にする．ひざや他の関節の損傷を診断する技術も利用でき，それは，内視鏡を膝関節に物理的に導入するアンスロスコピーに対して痛みのない代替法となる．

4. 水素に加えて，いくつかのタイプの原子が MRI で検出され得る．そして，^{31}P 原子に基づいた画像診断の応用が開発されつつある．この技術は，代謝の研究に大きな将来性をもっている．

[Ex. 44]　金　　属

1. 鉱石から金属の採取は，文明の発達と合致した。銅と錫の合金，青銅（ブロンズ）は広く用いられた最初の金属材料である。精錬技術が精巧になるとともに，鉄がより好まれる金属となった。青銅よりも硬い物質であり，剣や鋤により適していたからである。装飾用には，金や銀が使いやすかった。それらは非常に展性のある金属（すなわち，簡単に変形できる）であるためである。

2. 続く数世紀にわたって，知られる金属の数は，現在の大きな数，すなわち，周期表の元素の大部分に達した。それでも現代世界で，我々の生活を支配するのは，なお少数の金属，とくに，鉄，銅，アルミニウム，そして亜鉛である。我々の選ぶ金属は，それらを必要とする目的に合わなければならないが，それに加えて，鉱石が入手可能なこと，採取のコストが，しばしば，何故ある金属が他のものよりも選ばれるかの主な理由である。

3. SATP（標準周囲温度・圧力：25℃，100kPa）における高い三次元電気伝導度が，金属結合の 1 つの鍵となる特質であった。電子の共有がほとんどいつも不連続な分子単位内で起こる非金属と違って，金属原子は外殻（原子価）電子を金属構造全体にわたって共有する。このことが金属の

高い電気および熱の伝導性率，とともに，その高い反射性を説明するのに使われる。

4. 方向性をもった結合のないことは，金属原子がお互いに容易に滑りあって，新たな金属結合を作ることができるという点で，多くの金属の高い展性と延性を説明するのに使うことができる。金属結合の生成の容易さは，硬い金属を焼結できることを説明する；すなわち，我々は，鋳型に金属粉末を満たして，その粉末を高温高圧の条件に置くことで堅い金属の形を作り出すことができるのである。このような状況では，金属が実際に本体が融解することなく，金属-金属結合が粉末の粒界を通して形成される。

5. 単純な共有結合分子は一般に低い融点を持ち，イオン性化合物は高い融点を持つ一方で，金属は水銀の$-39℃$からタングステンの$+3410℃$まで広がる融点を持つ。金属は，融解状態でも熱や電気を伝導し続ける。(事実，融解アルカリ金属はしばしば原子力設備で熱移動媒体として用いられる。)これは，金属結合が液体状態でも維持されることの証明である。

6. 金属結合の強さと最も密接に関係するのは，沸点である。例えば，水銀は沸点$357℃$と原子化のエンタルピー $61\ kJ \cdot mol^{-1}$ を持つが，一方で，タングステンはそれぞれ $5,660℃$ と $837\ kJ \cdot mol^{-1}$ である。このように，水銀の金属結合はいくつかの分子間力と同じほど弱いが，一方，タングステンのそれは多重共有結合と強さが同じぐらいである。しかしながら，気相では，リチウムのような金属元素は対，Li_2 の形で，あるいはベリリウムのように個々の原子として存在し，したがって金属の性質を失う。気相の金属は金属に見えなくなることさえある；例えば，気相中カリウムは緑色を呈する。

7. 最も単純な金属結合モデルは，電子海（あるいは電子気体）モデルである。このモデルでは，原子価電子は自由に金属構造全体を動き（したがって「電子海」という名前），そして金属を離れ，そのため正のイオンを生成しさえする。さらにまた，電流を運ぶのは価電子であり，金属を通して熱を移動するのは価電子の運動である。しかし，このモデルは，定量的というよりは定性的である。分子軌道理論が金属結合のより包括的なモデルを提供する。この分子軌道理論の拡張は時にはバンド理論と呼ばれる。

[Ex. 45] 制 酸 剤

1. 店頭市販薬の主な範疇の1つは制酸剤である。事実，具合の悪くなった胃の治療は10億ドルのビジネスである。制酸剤は最も普通のタイプの無機薬品である。胃は酸—塩化水素酸—を含む。というのは，ヒドロニウムイオンは複雑なタンパク質を分解（加水分解）して胃壁から吸収されるような単純なペプチドにするための優秀な触媒であるからである。不幸なことに，いくらかの人々の胃は，酸を過剰に出す。過剰の酸の不快な効果を良くするには，塩基が必要である。しかし，塩基の選択は，化学実験室ほど単純ではない。例えば，水酸化ナトリウムの摂取は，ひどい，たぶん生命を脅かす喉の損傷を引き起こすであろう。

2. 具合の悪い胃に普通使われる薬はベーキングソーダつまり炭酸水素ナトリウムである。炭酸水素イオンは水素イオンと次のように反応する：$HCO_3^- + H^+ \rightarrow H_2O + CO_2$。この化合物には，1つの明らかな欠点，そして，もう1つのそれほど明らかではない欠点がある。その化合物は胃のpHを上げるが，また，ガスの生成（いわゆるおなかの張り）を導く。加えて，余分のナトリウムの摂取は高血圧の人には良くない。

3. いくつかの専売の制酸剤は，炭酸カルシウムを含む。これもまた，二酸化炭素を生じる：$CaCO_3 + 2H^+ \rightarrow Ca^{2+} + H_2O + CO_2$。人のカルシウム摂取を増すという有利な面は，このような制酸剤の配合物を売る企業が言及するが，カルシウムイオンが便秘剤として作用すること

にはめったに触れない。もう1つの人気のある制酸剤化合物は，水酸化マグネシウムである。これは錠剤の処方でも入手できるが，細かく砕いた固体を着色した水と混ぜて，マグネシアミルクと呼ばれるスラリーにしたものも市販されている。水酸化マグネシウムの低い溶解性は，そのサスペンション(懸濁液)中のフリーの水酸化物イオンの濃度が無視できることを意味する。胃の中で，不溶の塩基は酸と反応して，マグネシウムイオンの溶液を与える： $Mg(OH)_2 + 2H^+ \rightarrow Mg^{2+} + 2H_2O$。カルシウムイオンは便秘剤である一方，マグネシウムイオンは便通剤である。この理由で，いくつかの処方箋は炭酸カルシウムと水酸化マグネシウムの混合物を含んで，2つのイオンの効果をバランスさせる。

[Ex. 46] クラスレート，メタンと二酸化炭素ハイドレート

1. 数年前までは，クラスレートは実験室の珍奇なものであった。今は，メタンや二酸化炭素のクラスレートが，特に，大きな環境上の関心となりつつある。クラスレートは，他の分子の結晶の骨組みの中に分子あるいは原子が捕獲されている物質と定義される。ここでは，我々は水の気体クラスレートに焦点を当てよう。それは時々気体ハイドレート(水和物)と呼ばれている。後者の用語(ハイドレート)は広く使われるが，それは厳密には正しくない。ハイドレート(水和物)という用語は，通常，物質と周りの水分子の間で何らかの分子間相互作用を意味するからである。例えば，水和された金属イオンのように。

2. クラスレートを非常に重要な問題に換えたのは，海底上のメタンハイドレートの発見である。我々は，大面積の海洋の底には，堆積層の最上層の下にメタンクラスレートの厚い層があることを知っている。クラスレート層は，地質時代にわたって，(穴の開いた)下層の気体堆積床から上がってくるメタンと堆積層からしみ落ちてくる融点近くの水との相互作用によって，形成したというのがもっともらしい。ハイドレート各1立方センチメートルは，SATP(標準周囲温度圧力：298 K, 100 kPa)で約175 cm^{3}のメタンを含む。このクラスレートのメタン含量は，「氷」が実際に燃えるというのに十分である。世界の海のメタンクラスレート堆積床の全炭素量は，地上のすべての石炭，油，天然ガスの堆積の総和の全炭素量の2倍になると信じられている。

3. メタンクラスレートの安定性は，温度と圧力に大きく依存するので，海洋の温暖化がクラスレート堆積床の融解を導き，大量のメタンを大気中に放出するという心配がある。放出されたメタンは，メタンが有力な温室ガスであるので，気候に重大な影響を与えるであろう。何度かの急激な過去の気候変動はクラスレートからのメタン放出によって触発されたと議論されたことがある。例えば，氷河時代の水位の低下は海床堆積層の圧力を下げ，多分大容積のガスを放出したであろう。こうして，増加したメタン量が地球温暖化を引き起こして，氷河時代を終わらせたであろう。

4. 二酸化炭素の深海隔離は，発電所や産業過程で生成する廃棄二酸化炭素を貯蔵する1つの可能な方法として示唆されてきた。二酸化炭素が，周囲温度・圧力の条件下で，深海に放出されると，固体クラスレートを形成する。クラスレートは，高い安定性をもつ。例えば，250 mの深さで，圧力は2.7 MPaであり，クラスレートは+5℃で安定である。"正常な"氷は液体の水より密度が低いのに対して，二酸化炭素クラスレートは約1.1 g・cm^{-3}の密度をもち，海床に沈む。数メガトンもの過剰の二酸化炭素がこの方法で処理できると提案されてきた。

5. この概念と関連した3つの重要なことがある。最初の最も重要なのは，二酸化炭素クラスレートの層が，それが堆積される深海の珍しい海底の生命を窒息させるということである。第2に，実験ですでに示されたことで，魚が，実験的なクラスレート堆積物まわりの二酸化炭素飽和

水に近づくと呼吸困難を示すということである。第3に，長期間，多分数十万年にわたって，クラスレートはおそらくその包接した二酸化炭素を周囲の水に放出して，海洋のpH減少を引き起こす。pH変化は明らかに海洋生命のエコバランス(生態平衡)に影響を及ぼすであろう。

[Ex. 47] シリコン，シリカ，ゼオライト

1. 地球殻のおよそ27％質量はシリコン(珪素)である。しかし，シリコン自体は天然にフリーの元素としては見られなく，酸素−シリコン結合を含む化合物の形になっている。この元素は，灰色の，金属様の，結晶性固体である。金属のように見えるけれども，電気伝道度が低いので，金属としては分類されない。

2. 年間約50万トンのシリコンが金属の合金の調製に使われている。合金製造が主な用途であるが，シリコンはコンピューターを機能させる半導体として我々の生活に決定的な役割を演じる。電子産業で用いられるシリコンの純度レベルは非常に高くなければならない。例えば，たった1ppb ($1/10^9$) の燐の存在でも，十分，シリコンの比抵抗を150から$0.1\ \mathrm{k\Omega \cdot cm}$まで下げてしまう。高価な精製過程の結果として，超純度の電子グレードのシリコンは，冶金グレード(純度98％)シリコンの1,000倍を超える値段で売られている。

3. 普通にシリカと呼ばれる二酸化珪素SiO_2の最も普通の結晶形は，鉱物の石英である。多くの砂は，通常酸化鉄などの不純物を含むシリカの粒子から成る。二酸化炭素と二酸化珪素は同じタイプの分子式を共有するが，性質が非常に違うことに気づくと興味深い。二酸化炭素は室温で無色の気体であり，一方，固体の二酸化珪素は1,600℃で融解し，2,230℃で沸騰する。この違いは，結合の因子による。二酸化炭素は，小さい，3原子の，非極性分子単位から成り，その分子の互いの引力は分散力による。対照的に，二酸化珪素は，巨大な分子格子の中で珪素—酸素共有結合の網目を含んでいる。各々の珪素原子は4つの酸素原子に結合し，各々の酸素原子は2つの珪素原子に結合して，この化合物のSiO_2化学量論に一致した配置になる。

4. ちょっと考えると，アルミニウムという金属とシリコンという半金属/非金属は，共通するものはほとんどないと考えるであろう。しかし，大多数の鉱物の構造で，アルミニウムは部分的にシリコンを置き換えている。これは驚くことではない。というのは，アルミニウムとシリコンは同じサイズのカチオンサイトに適合するからである。もちろん，これは結合がイオン性であることを前提にしている。

5. 大多数のアルミノ珪酸塩は，実際，SiO_4単位が角の酸素原子で結合した三次元配列の基本的シリカ構造に由来する。シリカでは，その構造は中性である。したがって，Al^{3+}がSi^{4+}を置換すると，格子は1つの置換毎に1つの正味の負荷電を獲得することになる。例えば，珪素原子の1/4がアルミニウムで置換されると，経験式 $[AlSi_3O_8]^-$ のアニオンを生じる：珪素原子の半分を置換すると，式 $[Al_2Si_2O_8]^{2-}$ を与える。荷電は，1属または2属のカチオンで相殺される。この特定の組の鉱物は，花崗岩の成分である長石を構成する。典型的な例は，正長石 $K[AlSi_3O_8]$ と灰長石 $Ca[Al_2Si_2O_8]$ である。

6. ある三次元アルミノ珪酸塩構造は，網目全体を通した開いたチャンネル(溝)を持っている。この構造をもつ化合物はゼオライトとして知られ，その産業における重要性は急騰しつつある。多くのゼオライトが天然に存在するが，化学者は，構造全体を通して新たな穴を持ったゼオライトの大規模な探索を始めた。ゼオライトには主な4つの利用法がある：(1)イオン交換剤として，(2)吸着剤として，(3)気体吸収に，(4)産業用触媒として。

7. 最も重要な触媒の1つは$Na_3[(AlO_2)_3(SiO_2)] \cdot xH_2O$で，普通ZSM-5と呼ばれる。この

化合物は天然には生じなく，モービルオイルの研究化学者によって初めて合成された。これは，天然に存在する大抵のゼオライトよりもアルミニウム組成が高く，その機能を果す能力は，高電荷密度のアルミニウムイオンに結合した水分子の高い酸性に依っている。事実，ZSM-5の水素は，硫酸の水素と同じぐらい，ブレンステッド・ローリー酸として強い。ZSM-5は適当なサイズと形の分子をその穴の中に受け容れて，強酸として作用することによって，反応を触媒する。1例は，ベンゼンとエチレンから，重要な有機試薬であるエチルベンゼンの合成である。エチレンがまず，ゼオライト内でプロトン化されて，エチルカチオンを生じ，これがベンゼンを攻撃して生成物を与えると考えられている。

[Ex. 48] 身体の中のイオン類

1. 地球上天然にある90の元素の内，25が生きている生命有機体に必須である。あらゆる生命有機体にある主な元素は，水素，酸素，炭素，および窒素である——これらは，炭水化物，脂肪，タンパク質，およびビタミンのような有機化合物の中にある。残りの21の元素はミネラルである——これらは天然にある無機元素である。ミネラルは2つのグループ，主ミネラルと微量ミネラルに分けることができる。（訳註：mineralは通常「鉱物」と訳されるが，ここでは有機化合物中の元素に対比する総称の意味で「ミネラル」とした。）

2. 主ミネラル，あるいはマクロミネラルは，大量にあるもので，カルシウム，燐，カリウム，硫黄，塩化物，ナトリウム，およびマグネシウムを含む。カルシウムと燐は一緒になって，身体の中に存在するミネラルの質量の3/4を成す。微量ミネラルは，鉄，ヨウ化物，フッ化物，マンガン，亜鉛，モリブデン，銅，コバルト，クロム，セレン，砒素，ニッケル，珪素，ホウ素である。微量ミネラルは主ミネラルよりも少量に存在するとはいえ，それらも身体に同じように重要である。数マイクログラムのヨウ素の日常の不足は，数百ミリグラムのカルシウムの不足と同じぐらいに深刻である。

3. 元素のヨウ素は，通常天然にはヨウ化物イオン（I^-）で存在する。ヨウ化物イオンは身体に非常に少量だけ必要とされるが，この量を摂ることが重大である。身体は，チロキシン，すなわち基本代謝率を制御する原因となるホルモン，を作るのにヨウ化物イオンを利用する。ヨウ化物イオンがないと，身体はチロキシンを作れない。チロキシンは，下首に位置するリンパ腺である甲状腺上で合成される。血液のヨウ化物量が低いと，甲状腺細胞が拡大し，甲状腺腫を作る。この状態の人は無精と体重増を患う。この状態を患う母親に生まれる幼児は，クレチン病として知られる，取り返せない精神的・肉体的遅れをもって生まれるかもしれない。

4. カルシウムは，骨形成，歯形成，神経伝達，正常血圧の維持，血液凝固，筋肉収縮，および心臓機能に必須である。細胞は，常にカルシウムの摂取を必要とし，身体が血液中のカルシウムイオン濃度を維持するようにする。骨格は，そこから血液がカルシウムを借り，戻す，貯蔵所（銀行）の役をしている。適当なカルシウムを何年も摂らなくても，認められるような症状を患うこともなく過ごせるが，そうすると人生の後期に，カルシウムの貯蓄がなくなって，骨格の元のままの状態がもはや維持できないことが分かる。成人の骨の欠損，すなわち骨粗しょう症は多くの老人の健康問題である。毎年，米国の百万を越える人々が骨粗しょう症による骨折を患う。閉経は女性の骨損を増加させる。

5. 元素の燐は，主にリン酸化物イオン，PO_4^{3-}，として必要とされる。カルシウムイオンとともに，リン酸化物イオンは骨と歯の形成に必須である。リン酸化物は血液中のバッファー（緩衝液）の重要成分であり，血液の酸－塩基バランスを維持させる。リン化合物は，アデノシン3リ

ン酸エステル(ATP)を通して，細胞中のエネルギー移動に必須である．リンは，遺伝物質 DNA と RNA の必須の部分である．ある脂肪—リン脂質—は各細胞周りの膜を形成し，これもリンを含んでいる．

6. 鉄は，主に，ヘモグロビンとミオグロビンの成分として必要とされる．両方の化合物ともに，酸素を運び，保つのに鉄を利用する．ヘモグロビンは血液中の酸素運搬体であり，ミオグロビンは筋肉細胞中の酸素貯蔵所である．鉄はまた，エネルギーの代謝的移動に多くの酵素によって利用される．

7. バッファー（緩衝液）は，pH 変化に抵抗する溶液である．それらは，酸や塩基が溶液に加えられたときの大きな pH 変化から守る．身体の血液と細胞外流体はバッファーを含んでいる．あるシステムのバッファーとして働く能力は，巨大量の酸や塩基が加えられた時には，過重荷となるかも知れない；血液中のアシドーシス（酸性血症）またはアルカローシス（アルカリ性血症）を導くことになる．

8. リン酸2水素カリウム(KH_2PO_4)とリン酸水素カリウム(K_2HPO_4)の混合物は次の平衡に関わる：$H_2PO_4^- + H_2O = HPO_4^{2-} + H_3O^+$．このバッファー溶液に酸が加えられると，加えたヒドロニウムイオンからのストレスが平衡を左へ移動させ，このストレスを和らげる．余分のヒドロニウムイオンのほとんどは水に変えられ，pH は比較的一定のままである．塩基は溶液中のヒドロニウムイオンの濃度を減少させる．塩基がバッファー溶液に加えられると，減少したヒドロニウムイオン濃度から来るストレスが平衡を右に移動させる．水はヒドロニウムイオンに変換され，pH はあまり大きく変化しない．

[Ex. 49] 酸と塩基

1. 酸性と塩基性は電気陰性度と結合極性に関係がある．有機分子の酸—塩基挙動は，その化学の多くを説明する助けとなる．酸性の2つの定義がしばしば用いられる：ブレンステッド－ローリーの定義とルイスの定義である．

2. ブレンステッド－ローリー酸は水素イオン（すなわちプロトン H^+）を与える物質であり，ブレンステッド－ローリー塩基は水素イオン（すなわちプロトン H^+）を受ける物質である．酸は塩基と反応する：気体の塩化水素酸が水に溶けると，酸の HCl は塩基の水分子にプロトンを与える．この反応の生成物は新たな酸と塩基である．反応物から生成物を区別するため，生成物を共役酸および共役塩基と同定する．生成したヒドロニウムイオン(H_3O^+)は，プロトンを与え得るので，共役酸であり，塩化物イオン(Cl^-)はプロトンを受けることができるので，共役塩基である．

3. 酸がプロトンを与える能力は，酸に依存する．HCl のような強酸は，水とほとんど完全に反応するが，一方，酢酸(CH_3COOH)のような弱酸は少ししか反応しない．与えられた酸の水溶液中の正確な強さは，酸性度定数 K_a で表現することができる．一般化された酸 HA と水との反応に対して，その平衡と酸性度定数 K_a は，次のように書かれる：$HA + H_2O = A^- + H_3O^+$ ； $K_a = [H_3O^+][A^-]/[HA]$

ただし，括弧 [ ] は濃度を指し，カッコ内にいれた種のモル濃度すなわちリットル当たりのモル数(mol/L)である．より強い酸は，平衡を右側に向けるので，より大きな酸性度定数をもつ；より弱い酸は，平衡を左側に向け，より小さな酸性度定数をもつ．

4. さまざまな酸の K_a 値の範囲は膨大であり，最も強い酸の 10^{15} から最も弱いものの 10^{-60} までである．H_2SO_4, HNO_3, そして HCl のような普通の無機酸は 10^2 から 10^9 の範囲の K_a を持ち，

一方，多くの有機酸は 10^{-5} から 10^{-15} の範囲の K_a を持つ．もっと経験を得るにつれて，どの酸が「強く」，どの酸が「弱い」かという粗い感覚を持つようになるであろう（ただし，これらの言葉はいつも相対的なものであることを覚えておくこと）．

5. 酸の強度は，普通 pK_a 値を用いて与えられる．ここで，pK_a 値は K_a 値の常用対数の負の値に等しい：$pK_a = -\log K_a$

強い酸（より大きい K_a）は，より小さい pK_a を持ち，弱い酸（より小さい K_a）は，より大きい pK_a を持つ．いくつかの普通の酸は，強さの増す順に，次のような pK_a 値を持つ：水（H_2O）15.74，酢酸（CH_3CO_2H）4.76，フッ化水素酸（HF）3.45，塩化水素酸（HCl）-7.0

6. 酸・塩基のルイスの定義は，プロトンを与えたり，受けたりする物質に限らないという点で，ブレンステッド-ローリーの定義と異なる．ルイス酸は，空の原子価軌道を持つので，電子対を受け入れることができる物質であり，ルイス塩基は電子対を与える物質である．供与された電子対は，新たに生成した共有結合中でルイス酸と塩基の間に共有される：$A\ +\ :B\ \rightarrow\ A-B$

7. プロトン（H^+）は，ルイス酸である．なぜなら，塩基に結合するとき，電子対を受け入れて，その空の 1s 軌道を満たすからである．しかし，ルイス酸はプロトン供与体だけでなく，多くの他の種も含んでいる．3塩化アルミニウム（$AlCl_3$）のような化合物はルイス酸である．なぜなら，これも，トリエチルアミン（$N(CH_3)_3$）のようなルイス塩基から電子対を受け入れて，空の原子価軌道を満たすからである．

[Ex. 50] 超強酸

1. 超強酸は 100% 硫酸よりも強い酸と定義される．事実，化学者は硫酸の 10^7 から 10^{19} 倍強い超強酸を合成している．普通のブレンステッド超強酸は過塩素酸である．過塩素酸を硫酸と混ぜると，硫酸が塩基のように働く：$HClO_4 + H_2SO_4 \rightleftarrows H_3SO_4^+ + ClO_4^-$．フルオロ硫酸は最強のブレンステッド超強酸である：それは，硫酸よりも 1,000 倍以上酸性である．この超強酸は，-89℃ から +164℃ まで液体なので，理想的な溶媒である．

2. ブレンステッド-ルイス超強酸は，強力なルイス酸と強いブレンステッド超強酸の混合物である．最も強力な組み合わせは，フルオロ硫酸中 5フッ化アンチモン SbF_5 の 10% 溶液である．SbF_5 の添加は FSO_3H の酸性度を数千倍増加させる．2 つの酸の反応は非常に複雑であるが，混合物中の超水素イオン供与体は，$FSO_3H_2^+$ イオンである．この酸混合物は，普通の酸とは反応しないような多くの物質，例えば炭化水素と反応する．例えば，プロペン C_3H_6 はこのイオンと反応してプロピルカチオンを与える：$C_3H_6 + FSO_3H_2^+ \rightarrow C_3H_7^+ + FSO_3H$

3. FSO_3H 中 SbF_5 の溶液は通常「マジック・アシッド（魔法の酸）」と呼ばれる．この名前は，ケースウエスタンリザーブ大学・オラー（George Olah）の実験室に発している．オラーは超強酸分野のパイオニア（そして，1994 年ノーベル化学賞の受賞者）である．オラーのもとで研究していた研究員が実験室のパーティで残っていたクリスマスのろうそくの小片をこの酸に入れて，これがすばやく溶けたのを見つけた．彼は，生じた溶液を研究して，パラフィンワックスの炭化水素分子が水素イオンを付加して，生じたカチオンが自己転移して，分岐分子を生成したのを発見した．この予期されなかった発見が「マジック・アシッド〈魔法の酸〉」という名前を示唆したのであり，今はこの化合物の商標になっている．この一連の超強酸は石油産業で，あまり重要でない直鎖炭化水素から，もっと価値のある分岐分子への変換に用いられる．分岐分子はハイオクタンガソリンの生産に必要とされる．

[Ex. 51]　有機分子の性質

1. 2,700 万を超える有機化合物があり，その各々は独自の物理的，化学的性質を持っている。これらの化合物のいくらかは天然に存在する。他のものは化学者が創り出したものである。自然と人々が有機化合物を作るのに用いる「規則」は，主に，原子の小さい組み合わせから成る化学反応を理解することにある。そこで，ばらばらな反応性を持つ 2,700 万の化合物の代わりに，その化学を合理的に予想できるような数ダースの種族の化合物集団がある。我々は，最も広く知られている集団の化学を学ぶことにしよう。

2. 化合物を反応性で分類するのを可能にする構造的特長は，官能基と呼ばれる。官能基は，大きな分子内の原子のグループであり，特徴的な化学的挙動を持つものである。化学的には，ある与えられた官能基は，それが存在するあらゆる分子でほとんど同じように振舞う。例えば，最も単純な官能基は炭素―炭素二重結合である。炭素―炭素二重結合の電子構造はそれが存在するすべての分子で実質的には同じであるので，その化学反応性も同じままである。例えば，エチレンすなわち炭素―炭素二重結合を持つ最も単純な化合物は，ペパーミント油の中にあるかなり大きい分子であるメンテンの反応と同じ反応を行う。例えば，両者とも Br_2 と反応して，臭素原子が二重結合炭素のそれぞれに付加した生成物を与える。あらゆる有機分子の化学は，サイズと複雑さに無関係に，それが含む官能基によって決まる。

[Ex. 51]　アスピリン

1. 昔の原住のアメリカ人は，熱や痛みに抗するのに柳の木の皮を用いた。ヨーロッパ人が柳皮の薬性について知ったのは，1763 年，牧師のエドワード・ストーンがロンドンの王立学会への論文を読んだときであった。柳皮の抽出物は，結果的に，強力な鎮痛剤(痛みを和らげる)，解熱剤(熱を減じる)，および炎症抑制剤(腫れを減じる)の薬であることが分かった。有機化学者は 1838 年に，柳皮中の活性成分を単離し，同定することができた。成分のサルチル酸は柳の木のラテン語である salix から名づけられた。サルチル酸の使用は，その酸性が胃のひどい刺激を引き起こしたので，限られていた。

2. ドイツの化学者 フェリックス・ホフマン はバイエルの工場で働いていて，サルチル酸のエステル，アセチルサルチル酸を 1893 年に合成した。アセチルサリチル酸は，アスピリンの商標でバイエルによって市販された。世界第一次大戦の間，合衆国政府はバイエルの資産を差し押さえ，それらとバイエルの名称をスターリング・プロダクツ(現在，スターリング・ウインスロップ社)に売った。

3. サルチル酸と同様に，アスピリンは鎮痛剤，解熱剤，炎症抑制剤である；それは，サルチル酸よりも胃に対して，はるかに刺激性が小さかった。それは，頭痛，歯痛，風邪の痛みと熱，筋肉痛，月経痛，関節炎の痛みを和らげる。研究によると，1 日当たり少量のアスピリン 1 錠(アセチルサリチル酸 81 mg を含む)は，心臓発作，脳卒中，結腸がんのリスク(危険)を下げるかも知れないと，示唆されている。

4. アスピリンは僅かな胃腸の出血を引き起こし，これは，長時間では，鉄不足や胃潰瘍の原因となる。これらの併発症は，腸溶解性コートされたアスピリンで避けられる。これは，小腸に届くまでは溶解しない。アスピリンは，インフルエンザや水ぼうそうのある子供には，稀だがしばしば致命的となるライ症候群のリスクがあるので，与えるべきではない；これらの病気にかかっている子供は，アセトアミノフェンのような鎮痛剤を与えるべきである。医者に指示されないかぎり，アスピリンは妊娠後期 3 か月間に摂るべきではない。

5. やっとこの数年になって，生化学者は，アスピリンがどのようにその不思議なことをもたらしているのか理解し始めた。アスピリンは，プロスタグランジンの生成を阻止する。プロスタグランジンは，痛み，熱，局部的炎症を引き起こす原因となる一連の化合物である。多くの医薬品と同様に，アスピリンは天然由来の物質から開発された。化学者はまずその活性成分を単離し，その構造を決定し，それに次いでその原形を改良した。アスピリンとプロスタグランジンの相互作用の正確なメカニズムが解明されると，更なる改良が可能である。

[Ex. 53] 甘味料

1. 「砂糖」という言葉を言うと，大抵の人はすぐに甘い味のキャンディ，デザートのようなものを思い浮かべる。事実大抵の単純な炭水化物は甘い味がする，しかし，甘さの程度は1つ1つの糖で大きく違う。スクローズ(食卓砂糖)を標準とすると，フルクトースはほぼ2倍の甘さであるが，ラクトースはたった1/6の甘さである。しかし比較は難しい，というのは，甘さは味の問題であり，糖類のランク付けは個人的意見の問題であるから。

2. 多くの人々のカロリー摂取量を制限しようとする願いは，サッカリン，アスパルテーム，アセスルファムのような合成甘味料の開発を導いてきた。すべてが天然の糖類よりもはるかに甘く，180倍以上である。したがって，どれか1つの選択は，個人の味覚，政府の規制，そして(焼いたものでは)熱安定性に依存する。サッカリンは，最も古い合成甘味料で，幾分金属的なあと味があるけれども，1世紀以上も使われてきた。その安全性と発がん性の疑いが1970年代に生じたが，今は疑いが晴れている。アセスルファム・カリウムは，ごく最近認められた甘味料で，それはほとんどあと味がないので，ソフトドリンクに非常にポピュラーになっている。これら3つの合成甘味料はどれも，炭水化物には何の構造的類似性もない。

3. サッカリンは，最初の人工甘味料で，100年以上も前に発見された。普通の食卓砂糖の代替品を用いるのが流行となるよりも，ずっと前のことである。これは，19世紀の最も有名なアメリカの化学者 Ira Remsen の実験室でたまたま見つかったときのことである。

Remsen は1846年ニューヨークに生まれた；彼はドイツへ行って，ミュンヘン，ゲッチンゲン，チュービンゲンの大学で大学院研究を行った。合衆国へ戻って，彼はウイリアムズ・コレッジ，その後ジョン・ホプキンス大学の教授となった。彼は合衆国に，ヨーロッパに匹敵する質の高い最初の化学科を設立し，彼の学生の中に未来の多くの指導的アメリカ化学者を数えた。その後彼はジョン・ホプキンス大学の学長となった。Remsen の学生の一人は私の科学の「曽曽祖父」：E.P. Kohler であった。彼の学生の一人は James E. Conant で，その学生に Louis F. Fieser，その学生に Charles C. Price，その学生に Royston M. Roberts がいた。Roberts (本記事の著者)は，彼の化学の祖先を有機化学の父であるヴェーラーに辿ることができる，と好んで指摘した。というのは，Remsen は Rudolph Fittig の学生であり，Fittig は Friedrich Wöhler (ヴェーラー) の学生であったからである。

4. 1879年，Remsen の仕事仲間の一人は，進行中の理論的研究計画の一部として指定されていた問題を追及していた。これをしている間に，Fahrberg という名前のこの仲間は，彼が合成し，偶然自分の手にこぼした物質が異常に甘い味がしたことに気づいた。(化学者は，当時，自分の研究している物質の匂いをかいだり，味をみたりすることに，今ほど注意をしなかった。) Fahrberg は新しい甘味物質の可能な重要性を予見していたと思われる，というのは，商業的な合成法を開発し，1885年それについての特許をとったからである。彼がそれに対して選んだ名前は，糖に対するラテン語 saccharum から，サッカリンであった。

5. アスパルテームは，化学的には，ジペプチド，L-アスパルチル-L—フェニルアラニンのメチルエステルである。これも，化学者 James M. Schlatter によってまったく偶然で強い甘味料であることが分かった。アスパルテームの甘味は，成分のアミノ酸の性質の知識からは予想され得なかった。—(というのは)一方の成分は「まずい」味を持ち，他方は苦いからである。極度に甘い味は，2つの(アミノ酸の)組み合わせとメチルエステルへの変換から生じたもので，まったくの驚きであった。

Schlatter は自分の本(1984)の中で実際の発見について次のように述べている。「1965年12月，アスパルテームをメタノールとフラスコで温めていた，そのとき混合物がフラスコの外に飛び出た。その結果，いくらかの粉が指についた。少し後になって，1枚の紙を取り上げるために自分の指をなめたときに，非常に強い甘い味に気がついた……。」そのまま排泄されるサッカリンと違って，アスパルテームはその成分の天然アミノ酸まで代謝され，これはさらに通常の身体経路で代謝される。Schlatter はペプチドの代謝についてこのことを熟知していたので，彼は大胆にもフラスコの外にはね飛んだ物質の味をみたのである。

[Ex. 54] テルペン類

1. 多くの植物原料を蒸気と一緒に蒸留(水蒸気蒸留)すると，精油と呼ばれる香りの高い液体混合物が生産されることは，何世紀もの間知られていた。数百年にわたって，このような植物抽出物は，薬，スパイス，香料として使われてきた。精油の研究は，また，19世紀の科学としての有機化学の出現に大きな役割を演じた。

2. 化学的には，植物精油は主にテルペン—膨大な多様性の構造をもった小有機分子—と呼ばれる化合物の混合物からできている。数千のさまざまなテルペンが知られていて，多くは炭素—炭素二重結合をもつ。いくつかは炭化水素であり，他は酸素を含む；いくつかは開鎖分子で，他は環を含む。例えば，

3. すべてのテルペンは，その見かけの構造の違いに関わらず，関連している。イソプレン則と呼ばれる形式によると，テルペンは5炭素のイソプレン(2-メチルブタ-1,3-ジエン)単位の頭—尾結合から生じたと考えることができる。炭素1はイソプレン単位の頭で，炭素4が尾である。例えば，ミルセンは2つのイソプレン単位が頭—尾に結合して，2つの1炭素分岐を持った8炭素鎖を形作る。α-ピネンは同様に，2つのイソプレン単位を含み，もっと複雑な環状構造に集合している。

4. テルペン類は，それらが含むイソプレン単位の数によって分類される。すなわち，モノテルペンは2つのイソプレンから成る10炭素の物質，セスキテルペンは3つのイソプレンからの15炭素分子，ジテルペンは4つのイソプレン単位からの20炭素の物質などである。モノテルペンとセスキテルペンは主に植物中に見られるが，高級テルペンは植物と動物両方にあり，多くは重要な生物学的な役割をもっている。例えば，トリテルペンのラノステロールはステロイドホルモンの前駆体である。

5. イソプレン自体はテルペン類の真の生物学的前駆体ではないことが研究によって明らかになった。代わりに，自然は，2つの「イソプレン等価体」—イソペンテニル2燐酸とジメチルアリル2燐酸—を用いる。それらは5炭素分子で，自体は酢酸から作られる。酢酸からラノステロールを経て人のステロイドへの生物学的変換のそれぞれの段階が解かれている—いくつかのノーベル賞が授与されている膨大な偉業である。

[Ex. 55] 医　薬

1. 薬はどこから来るのだろうか。この質問に対する答えは，誰に尋ねるかによって，「陸」，「海」，「きつい仕事」であったり，「運」であったりするだろう。歴史的には，薬の元は薬用植物である。ヒポクラテスは，歯痛を治すのに，柳の皮を噛むことを勧めた。2000年以上も後になって，ドイツのバイエル社の化学者は，実験室で，アスピリン，すなわち柳の木の皮中にあるのと同じ活性成分—サルチル酸の誘導体を合成した。

2. 実際，それがマラリヤの薬であるか，今一番注目の新がん治療薬かに関わらず，自然はしばしばその先導役（リード）を提供する。事実，1981年から2002年の間に米国食品・薬品局に報告された1,031個のリード新薬の67％が，米国がん研究所主導の研究によれば，自然由来であった。（しかし，すぐ分かるように，これらの分子のほんの少しの部分が医薬となる。）

3. これらの薬の出所は，次のように4つの部分で表される。第1（15％）は，比較的新しく成長している出所として，ワクチン，タンパク質およびペプチド薬が遺伝子工学のバクテリアあるいは他の微生物を用いた発酵で生産される。第2の部分（28％）は，天然で見られるのとまったくそのままで用いられる分子である。経済的な理由で，これらの分子は天然で生産されてその後単離されるか，あるいは，実験室で合成的に作られるかである。第3の部分（24％）は，自然には存在しない分子であるが，分子の活性部分は天然由来で，非天然分子中に取り込まれている。第4（33％）は，天然源はなく，化学者が実験室で合成した分子からくる。

4. すべてこれらの化合物源があるのに，なぜ，医薬はそれほど遅く市場に出て，出たときにはそれほど高価なのか。答えは時間である。科学者がある可能性のある薬をいったん同定したとして，その薬が市場に出るまでに平均して11年を要する。これらの11年はいろいろな段階に分けられる。前臨床的展開は，主に，科学的および経済的問題に焦点がある。薬は，動物に安全で，人の（臨床）試験に必要な大規模で生産するのに十分安価でなければならない。

5. 臨床試験の第1段階では，薬は健康な人に投与され，副作用についてモニター（監視）される。第2段階では，その薬が改良し，治すと期待されるような病気あるいは状況をもった比較的少数の患者に，薬が投与される。第3段階の研究では，より広い範囲の患者にその薬を摂らせる。もしこれらの研究が成功すれば，会社は広範囲の臨床利用についてFDAからの承認を要請する。新たに承認された薬は，会社に数十億ドルの価値があり得る。

6. 第3段階でも，成功した製品を保障するものではない。時々，少人数の患者が，その薬が数年間広汎に普及した後で，不快で深刻な副作用を生じ，会社が市場からその薬を回収する結果となる。加えて，薬はいつも十分な利益を生むとは限らない。1つの理由は，競合者が常に新薬を導入することである。もう1つの理由は，これらの薬を保護する特許は20年間だけ有効であることである。特許は前臨床試験の間に提出されるので，会社は約6年だけその薬の占有権をもっているにすぎない。それ以後になると，ジェネリック薬品製造会社が市場に入ってきて，その薬をかなり安い料金で売ることになる。なぜなら，これらの会社は前臨床試験や臨床試験に関係した費用は要らないからである。

[Ex. 56] アルカロイド

1. アルカロイドは，顕著な生理活性を持った，天然由来の窒素化合物である。他の単純なアミンと同様，アルカロイドは塩基である；アルカロイドの名前はそのアルカリ性からきている。これらの複雑な分子は植物から得られ，多くは医薬用途をもっている。それらは一般に水に不溶であり，しばしば，市販の製品では水溶性の酸塩の形で見られる：

$$\text{アルカロイド(水不溶性)} + \text{酸} \longrightarrow \text{アルカロイド酸塩(水溶性)}$$

純粋なアルカロイドは，その酸塩を水酸化ナトリウムのような塩基で処理して，沈殿として得られる：

$$\text{アルカロイド酸塩(水溶液)} + \text{塩基} \longrightarrow \text{アルカロイド(水不溶性)} \downarrow$$

2. よく知られたアルカロイドにはカフェインとキニンがある。カフェインはコーヒー，茶，コーラ豆中にある。カフェインは中枢神経系に対する穏やかな刺激剤で，警戒心の増大と眠気を取り去る能力を引き起こす。コーヒーは2〜5%のカフェインを含み，茶葉中にあるのとおよそ同量である。多くの人々はカフェイン抜き(デカフェ)コーヒーを飲むのを好む。カフェインは，アルカロイドをロースト(焙煎)する前に，安全な溶媒として二酸化炭素液体(訳注：超臨界流体)を利用して，豆から抽出される。

3. コーラ・ソフトドリンク飲料は，カフェインを含むコーラ豆の抽出物から作られる。製造者は，リン酸，カラメル，甘味料，と炭酸水を加える。彼らは通常，コーラ抽出物からすべてのカフェインを取り除き，後で，米国食品・薬品局(FDA)の要求する正確な量を加えている。カフェインはまた，いくつかのエネルギー飲料にも加えられている。これらはFDA規制に含まれないので，コーラやソーダ飲料よりも多いカフェイン量があるかもしれない。

4. カフェインは，少量なら血圧に影響はないが，大量は血圧を上げる。コーヒー，茶やコーラ飲料を大量に飲む人は，カフェインに対する抗薬力とそれへの依存性の両方を発現し得る。大量使用者は，18時間の禁欲のあと，無気力，頭痛，さらには吐き気といった禁断症状を経験し得る。

5. カフェインの構造は，アデニンやグアニン(遺伝子物質DNAの重要な成分)の構造に非常に似ている。この類似性は，カフェインががんや出生欠陥の原因になるかもしれないという懸念を引き起こした。今のところ，この懸念を支持する証拠はほとんどない。ある人々はカフェインが習慣性の薬であると考え，いくつかの宗教はカフェインを含む飲料の利用を禁じている。

6. キニーネは解熱剤(熱を減じる薬)で，長い間，マラリヤに対する唯一の知られている治療薬であった。キニーネは，マラリヤに感染した細胞のDNAに結合して，その複製を禁ずる。感染された細胞だけが影響を受ける，というのは，それらは感染していない細胞よりも高濃度でキニーネを吸収するためである。このアルカロイドは，キナの木の皮の中に見つけられる；これらの木は19世紀後半インドネシアで広範囲に栽培された。

7. 世界第2次大戦で，日本のインドネシア侵略が連合国軍の必要としたキニーネの供給を断ちきったとき，アメリカの化学者，ロバート・ウッドワードはコールタールからキニーネを合成することに成功した。ウッドワードは複雑な有機物質を合成する能力で著名であり，1956年のノーベル化学賞を授与された。大抵のアルカロイドは，苦い味があり，キニーネはしばしば，味の研究で苦味の標準参照物として用いられる。トニック水(炭酸水)の苦い味はキニーネのせいである。キニーネの水溶液は高い蛍光性があり，紫外光(黒光(訳注：紫外線のこと))の存在下，薄い青色に見える。

[Ex. 57] 化学の橋でバイオマスと石油加工を繋ぐ

1. 石油は液体燃料の主な原料であるだけでなく，我々の用いるほとんどの化学品や高分子物質の主成分となるものである。石油供給の寿命や安定性についての永続的な疑問は，その生産と利用についての環境問題と同様，動物あるいは木材生物資源のような代替物の開発を駆り立ててきた。もし，炭水化物が，石油と同じように反応する，酸素化度の少ないグループをもった化合

物に変換することができたら，それらは，石油化学産業プラントに再生可能な原料を供給できることになる。

2. Bond らおよび Lange らによる最近の2つの報告は，グルコースなどの単糖の脱水生成物であるレブリン酸（LA）がこの要求を満たすことを示唆している。LA の中間体としての導入は，化学産業の構造基盤と完全に両立できる触媒的変換プロセスの利用を可能にする。

3. 糖の脱水は，酸を用いた処理で行われ，最終的には LA とギ酸を約3：1の重量比で生成する。この変換は何十年も知られていたが，通常は扱いにくい副生成物の生成を伴うので，混合物から LA を分離するのが困難である。バイオリファイナリー・プラットフォーム化学品としての LA の地位は，1990年代初期に Biofine Renewables 社によって，そして，LA の高収率での生産を可能にした2反応器システムの開発によって，実質的に格上げされた。

4. LA の燃料への変換は，LA の γ-ブチロラクトンへのよく確立されている水素化で始まる。Lange らは，この変換を，50を越す水素化触媒をテストすることで，最適化した。LA の，触媒（酸化チタン上に分散させた1重量％のプラチナ（白金）金属）存在下，40-bar（10^5Pa）H_2 を用いた200℃での還元は，95％変換効率（反応した出発原料の割合）で進行した。

5. Bond らは，水溶液中 GVL から CO_2 を触媒的に脱離して，ブテン異性体の最初の混合物を与える2段階プロセスを述べている。2個以上のブテン単位がオリゴマー化して，輸送燃料として使える十分な分子量の高級炭化水素（8個以上の炭素原子）を創り出す。

6. Lange らは，LA を水素化して，バレリン酸（VA）を主生成物として生成し，これをアルコールと反応させて，ガソリンあるいはディーゼル燃料添加物として適当な一連のバレリン酸エステルを生成させる。低分子量エステル（バレリン酸メチル，エチル，およびプロピル）は，10および20％（体積）レベルのガソリン添加物としての使用に適した性能を示した。高級エステル（バレリン酸ブチルおよびペンチル）は，直接ディーゼル燃料あるいはディーゼル添加物として使うことができた。レギュラーガソリンに15％バレリン酸エチルのブレンドを用いた道路試験が10車両，トータル約250,000 km の運転で行われた。エンジン性能の問題は何も認められなかったが，バレリン酸エチルのエネルギー密度の低さが，体積当たりの燃費の損失という予想された結果となった。

7. このように，機能的に有用なバイオレファイナリー中間体は，現在石油化学産業で用いられている化合物と構造的に同じである必要はない。十分に安全が確保され，実証済みとなっている生物燃料（エタノール）を，有望には見えるがまだそれほど認められていない代替物で置き換えられるには，その前に技術開発の問題が処理されなければならない。研究のチャレンジもある：最初の糖類から LA の生産も今なおそれほど高効率ではなく，燃料生産というシナリオ（筋書き）での生育する可能性を確かにするのに必要なスケールではいまだテストされていない。

8. 最後に，避けられない LA 副生成物として生成するギ酸をどうするかという疑問が処理されねばならない。同様な状況はバイオディーゼル産業にも存在する，ここではグリセリンという共生成物の生成を処理しなければならない。ギ酸は，触媒的転移水素化を経て LA を GVL へ還元するのに利用できるが，このプロセスはいまだ最適化されていない。それでも，生物ベースの化学薬品を石油化学技術と両立し得るプロセスの中間体として実証することは，次世代の生物燃料源として，さらには従来の原材料供給の代替品としての，再生可能な生物資源の利用に拍車をかけるであろう。

[Ex. 58] アミノ酸からタンパク質

1. タンパク質は，あらゆる生物体に存在する大きな生体分子である。それらは色んなタイプがあり，多くの生物学的機能を持っている。皮膚や指爪のケラチン，体のグルコース代謝を調節するインシュリン，細胞中 DNA の合成の触媒となる DNA ポリメラーゼは，すべてタンパク質である。その見かけや機能に関わらず，すべてのタンパク質は化学的に類似している。すべてが多くのアミノ酸から成り，これらが互いにアミド結合によって長鎖に繋がれている。

2. アミノ酸は，名前が示すように，2官能性である。それらは，塩基性のアミノ基と酸性のカルボキシル基の両方を含む。タンパク質のビルディングブロック（構成単位）としてのそれらの価値は，1つのアミノ酸の $-NH_2$ が別のアミノ酸の $-CO_2H$ の間にアミド結合を作ることによって，アミノ酸が結合しあって長い鎖になることができるという事実から生じる。分類の目的から，50個より少ないアミノ酸を持つ鎖はペプチドと呼ばれ，一方，タンパク質という言葉は，もっと長い鎖に用いられる。

3. アミノ酸は酸性と塩基性両方のグループを含むので，それらは分子内の酸—塩基反応を行い，主に双極性イオンすなわちツヴィッターイオン（両性イオン：ドイツ語の zwitter は「混成」を意味する）の形で存在する。

アミノ酸両性イオンは塩であり，したがって，塩と関係した多くの物理的性質を持っている。それらは，水に溶けるが，炭化水素には不溶，高融点をもつ結晶性の物質である。加えて，アミノ酸は両性である：すなわち，周囲の状況に応じて，それらは酸あるいは塩基として，どちらでも反応することができる。酸性水溶液中では，アミノ酸両性イオンは塩基であり，プロトンを受容してカチオンを生じ，塩基水溶液中では，両性イオンは酸であり，プロトンを失って，アニオンを生じる。

[Ex. 59] 脂　　質

1. 脂質は，小さい天然由来の分子で，水への溶解性はわずかで，有機体から非極性有機溶媒を用いた抽出によって単離することができる。脂肪，油，ワックス（ろう），多くのビタミンとホルモン，そして大抵の非タンパク質の細胞膜成分が，（具体）例である。この定義は，脂質が構造ではなくて，物理的性質（溶解性）で定義されているという点で，炭水化物やタンパク質で用いられた（定義の）種類と違っていることに，気付きなさい。

2. 脂質は，2つの一般的なタイプに分類される：すなわち脂肪やワックスのように，エステル結合を含んでいるので，加水分解できるものと，コレステロールおよび他のステロイドのように，エステル結合を含まず，加水分解できないものである。蜜蝋は，例えば，エステル結合を持つ構造 $[CH_3(CH_2)_2CO_2(CH_2)_{27}CH_3]$ の脂質を含んでいて，加水分解で，対応する酸とアルコール $[CH_3(CH_2)_2CO_2H$ と $HO(CH_2)_{27}CH_3]$ を生じる。

3. 動物の脂肪および植物の油は，最も広く存在する脂質である。それらは違っているように見える—バターやラードのような動物脂肪は固体であるが，一方，コーンオイルやピーナッツオイルのような植物油は液体である—けれども，それらの構造は密接に関係している。化学的には，脂肪と油はトリアシルグリセリン（トリグリコシドとも呼ばれる），すなわち，グリセリンと3つの長鎖カルボン酸とのトリエステルである。脂肪あるいは油を NaOH 水溶液で加水分解すると，グリセリンと3つの長鎖脂肪酸のナトリウム塩を生じる。

4. 得られる脂肪酸は，一般に直鎖であり，12 から 20 の偶数個の炭素原子を含む。1つ以上の二重結合が存在すると，それらは通常 Z（cis）幾何異性を有する。ある特定の分子の3つの脂肪

酸は同じである必要はなく，与えられた源からの脂肪あるいは油は，多くの異なるトリアシルグリセリンの複雑な混合物である可能性が高い。

5. 100を超えるさまざまな脂肪酸が知られていて，約40個が広く存在している。パルミチン酸(C_{16})とステアリン酸(C_{18})はもっとも豊富にある飽和脂肪酸で，一方，オレイン酸，リノール酸，リノレン酸(すべてC_{18})は最も豊富な不飽和酸である。オレイン酸は1個だけの二重結合をもつので，モノ(単)不飽和物であり，一方，リノール酸とリノレン酸は1個を超える，すなわちそれぞれ2および3個の二重結合をもつので，ポリ(多)不飽和物である。

6. 石けんは少なくとも紀元前600年，フェネキア人がヤギの脂肪を木灰の抽出液とともに沸騰することで凝乳状の物質を作ったときから知られている。しかし，石けんの洗浄性は一般には知られていなくて，石けんとしての利用は18世紀になるまで広がらなかった。化学的には，石けんは，上記のように，アルカリを用いた脂肪の加水分解(けん化)により作られる長鎖脂肪酸のナトリウムまたはカリウム塩の混合物である。

7. 生の石けんの凝乳は，石けんとグリセリンと過剰のアルカリを含んでいるが，水とともに沸騰させ，NaClを加えて精製することができ，純粋なカルボン酸ナトリウム塩を沈殿させる。生じる滑らかな石けんは乾燥し，香料をつけ，棒状に圧縮される。色つき石けんには染料が加えられ，薬用石けんには防腐剤が加えられ，磨き石けんには軽石が加えられ，浮く石けんには空気が吹き込まれる。

8. 石けんは，石けん分子の2つの末端が非常に違っているために，洗剤として作用する。長鎖分子のカルボキシレート末端はイオン性で，したがって親水性(水を愛する)である。結果として，それは水に溶けようとする。しかし，分子の長い脂肪族鎖部分は非極性で，疎水性(水を恐れる)である。それは水を避けて，グリースに溶けようとする。これら2つの相反する傾向の正味の結果は，石けんがグリースにも，水にも引きつけられ，そのために洗剤として有用になる，ということである。

9. 石けん分子が水中に分散あるいは溶解されると，長い炭化水素テール(尾)は群がって疎水性の球になる。一方，イオン性のヘッド(頭)は水相に突き出る。これらの球状の塊は，ミセルと呼ばれ，様式図のように示される。グリースや油滴はミセル中心の石けん分子の炭化水素非極性テールによって包まれて，水中に可溶化される。いったん可溶化されると，グリースと汚れはすすぎ出される。

10. 石けんは，それがなかった場合よりも，生活をずっと快適なものにしてきたが，欠点もある。硬水中では，金属イオンを含むので，可溶性のカルボン酸ナトリウム塩は不溶性のカルシウムやマグネシウム塩に変えられ，バスタブ周りにかすや，衣類に灰色がかったものを残す。化学者は，長鎖アルキルベンゼンスルホン酸の塩に基づいた一連の合成洗剤を合成することによって，これらの問題を免れた。石けんと違って，スルホン酸塩は硬水中で不溶な金属塩を作らず，不快なかすを残すこともない。

11. ステロイド類は，下記のコレステロールで示したような，特徴的な4環炭素骨格をもった植物および動物の脂質である。ステロイドは，体組織中に広く存在し，多くのいろんな種類の生理活性を有する。もっとも重要なステロイドの中には，コレステロールや性ホルモン(アンドロゲンとエストロゲン)，および副腎皮質ホルモンがある。

[Ex. 60] 炭水化物の消化

1. 炭水化物は，ポリヒドロキシアルデヒド，ポリヒドロキシケトン，および加水分解で分解

できてポリヒドロキシ単位を生じる大きな分子を含む，化合物群である．モノサッカライドはそれよりも小さい単位に分解できない．モノサッカライドすなわち単糖は，分子中の炭素数および構造中に見られるカルボニル官能基の種類によって，分類される．すなわち，グルコースはアルドヘキソース（6個の炭素原子から成るアルデヒド）であり，フルクトースはケトヘキソース（6炭素原子のケトン）である．

2. 身体の細胞が食事の炭水化物中に蓄えられているエネルギーを利用できる前に，炭水化物は消化・吸収されなければならない．消化は，複雑な分子が単純な分子に分解される過程である．これらの単純な分子は，吸収の間に，腸壁を通して血流に入る．吸収は透析の1つの形，すなわち，膜を通しての小分子の移動である．

3. 炭水化物の消化は口の中で，歯が食物を小片に噛み砕いたときに始まる；より小さな小片は大きい表面積をもち，より早く消化される．唾液は酵素（アミラーゼ）を含み，これがでんぷん（大きい多糖）の加水分解を始めて，デキストリン（小さい多糖）やマルトース（二糖）にする．飲み込んだ後で，食物は胃に入り，ここでタンパク質と脂肪の消化が始まるが，炭水化物の消化は一時中止される；お腹の胃液の低いpHが唾液の酵素を不活性化するのである．

4. 食物が小腸中に入ると，アルカリ性の膵液および腸液によって中和される．これらの液は，また，複雑な炭水化物の加水分解を再開する酵素を含んでいる．最終的に，すべての多糖および二糖はグルコース，フルクトース，およびガラクトースに分解される．これらの単糖はサイズが十分小さく，腸壁を通過して，血液中に吸収される．血液中を循環した後で，フルクトースとガラクトースは肝臓によってグルコースに変換される．血液中のグルコースはただちに細胞活動のエネルギーを供給するのに使われるか，あるいは，肝臓や筋肉中でグリコーゲンとして蓄えられる．

5. グルコースの水溶液は3つの形の平衡である：α-グルコース，β-グルコース，そして開鎖グルコースである．開鎖構造は平衡混合物のほんの0.02 %を占めるだけである．グルコース水溶液は36%がα-構造，64%がβ-構造で存在する．α- およびβ-グルコースは環状構造で，5個の炭素原子と1個の酸素原子が6辺の環を形成する．6辺のグルコース環は平面ではなく，ひだ状の「いす形」配座で存在する．α- およびβ-形の間で唯一の違いは炭素1上のヒドロキシル基の位置である．ヒドロキシル基が下を向く（アキシャル）なら，それはα-形である．ヒドロキシル基が上を向く（エカトリアル）なら，β-形である．

[Ex. 61] 核酸と遺伝

1. 生物体の遺伝情報は，DNA鎖中に糸に繋がったデオキシリボヌクレオチドの連鎖として蓄えられている．情報が保存され，未来の世代へ伝えられるためには，DNAをコピーするためのある仕組みが存在しなければならない．情報が利用されるためには，DNAメッセージを解読し，それが含む指示を実行するための仕組みが存在しなければならない．

2. クリックが「分子遺伝子学の中心ドグマ（定説）」と呼んだものは，「DNAの機能は，情報を蓄え，それをRNAに伝えること」と言う．RNAの機能は，引き続いて，DNAから受け取った情報を読み，解読し，利用して，タンパク質を作ることである．DNAの正しい部分を正しいときに解読することによって，生物体は遺伝子情報を利用して，機能するのに必要な何千ものタンパク質を合成する．

3. 3つの基本的な過程が，遺伝子情報の伝達の中で起こる：複写は，遺伝子情報が保存され，次世代に受け渡されるように，DNAの同じコピーが作られる過程である．転写は，遺伝子の

メッセージが読まれ，細胞核から，タンパク質合成の起こるリボゾームに運び出される過程である。翻訳は，遺伝子メッセージが解読され，タンパク質の合成に用いられる過程である。

[Ex. 62]　ポリエチレンとポリプロピレン

1. エチレンとプロピレンは，二重結合を持った多くの他の分子の中で(とくに)，「付加重合体(または付加ポリマー，付加高分子)」と呼ばれるものを生成する。このことは，重合可能なアルケンのどの原子もポリマーの生成のときに失われないということを意味する。数千ものエチレンおよびプロピレンモノマーが結合しあって，それぞれ，ポリエチレンおよびポリプロピレンを生成する。

2. 付加の機構(仕組み)は別にして，この全体の過程は，アルケンのπ-結合をσ-結合に変換させて，ポリマー中に繰り返し単位を結合させるということを伴う。σ-結合はモノマー単位中の切断されるπ-結合よりもはるかに強いので，多大のエネルギーが放出され，熱力学的表示は大きな負のエンタルピー，ΔHを伴う，すなわち重合に有利であり，重合は発熱反応となる。

　　訳注：4の付注から，形式的平均でσ-結合の解離エネルギー$= 250$ kJ mol^{-1}，π-結合の解離エネルギー$= 170$ kJ mol^{-1}

3. 他方，別々に自由に動いていた多くのモノマー分子がポリマー鎖中に結合し合うことは，この過程の間にエントロピーの大きな減少をもたらす，その結果，ΔSは負となり，重合に不利となる。(つまり)重合は秩序化の過程なのである。

4. これら2つの競合する熱力学的因子がすべての重合を支配する。そしてエチレンとプロピレンでは，重合温度で，エンタルピー項がエントロピー項を凌駕する(超える)。もしこれが事実でないとすると，ポリマーは生成されない，そして事実これら2つの競合するエンタルピーとエントロピー因子のゆえにすべてのポリマーはある温度以下でのみ生成され得る。この温度は天井温度として知られている。

5. フリーラジカル源をエチレンの気体試料へ加えると，白色粉末，すなわちポリエチレンを急速に生成し，それは融解されて，非常に多くの見慣れた商品に成形できる。一方で，このようなラジカルのプロピレンへの付加は短い炭化水素の役に立たないのり状物質を生成した。鎖の成長がうまくいく前に止まってしまったのである。上にエチレンで概説した熱力学的表示はプロピレンにも当てはまるにも関わらずである：ポリプロピレンの形成に対して熱力学の結論からは何の障害もないのである。この事実はアリル共鳴によってうまく説明されている：

　すなわち，プロピレンから共鳴安定化したアリルラジカルを生成する可能性が，フリーラジカル重合に立ちはだかるのである。

[Ex. 63]　ゴム弾性

多くのポリマーと同様に，天然ゴム中のポリマー鎖は，非常に長く，ランダムな(無秩序な)コイルとなった形でリラックスした(ゆるんだ)状態で存在している。ランダムコイルを思い浮かべるベストの方法は，三次元でのランダムな経路を辿ることを考えてみることである。これは，とても無秩序な並び方であり，一皿の料理したスパゲッティにたとえられている。ゴムを引っ張ると，個々のポリマー鎖は，無理やり，もっとはるかに延びた形に変えられる。この応力を取り除くと，鎖は直ちにもとのランダムコイルにもどる。あるポリマーが弾性体(ゴム)となるには，わずかなエネルギーの変化でその形をたやすく可逆的に変化できることが必要である。このように，ゴム弾性は，実質的にコンフォメーション(立体配座)変化から来るもので，いわゆるエント

ロピー弾性である：延伸と弛緩は，それぞれ，エントロピーの損失と取得を伴うからである。
(内容については，Ex. 22 も参照。)

[Ex. 64] シリコーン類

1. シリコーン類は，その骨格が珪素と酸素原子が交互になった長い，柔軟な鎖のポリマーである。骨格からぶら下がって，ブレスレットの飾りのように，側鎖があり，通常小さく，炭素ベースの単位であり，これらの側鎖の選択によってシリコーンに目覚しい広範囲の性質が得られる。撥水性のシリコーングリースは，メチル基のような油性の非極性側鎖をもっている。非極性側鎖と極性の水分子は混ざらないので，シリコーンから水をはじく。

2. シリコーン類の別の応用は，極性と非極性側鎖の注意深いバランスによる。少量のシリコーン発泡剤はポリウレタンフォームの泡のサイズを制御する。高比率の極性側鎖は，泡をより泡立つようにする。泡は大きくなり，開いた気孔を作って，車のシートや家具のクッションにあるような柔らかいフォームを生産する。極性の側鎖の数を減らすと，泡は小さくなる。これらの小さい泡は開かずに気孔を形成し，フォームははるかに硬い固体となって絶縁に用いられる。

3. シリコーン類は，ほかにも，見かけは矛盾した性質をもつ。パン屋のパン鍋をコートしているシリコーン樹脂は，新しく焼いたパンが鍋にくっつくのを防ぎ，工場の鋳型につけた液状シリコーンポリマーは新しく作ったタイヤに同じことをする。しかし，「粘着付与用樹脂」を加えて，シリコーンをべとべとにして，ニコチン(タバコを止めようとしているスモーカー用)やスコポールアミン(船酔いに悩んで，これをなくさないようにしている人用)が入っている皮膚用絆創膏に用いる薬物透過性の接触粘着剤を生産する。

4. 珪素と酸素は地球上に最も豊富な2つの元素であり，それらは天然で結合して，珪酸塩を形成する。これは，ガラスや，石英，花崗岩のような物質を含む。これら2つの元素は1930年代合衆国で，シリコーンとして，初めて合成的に結合された。それらはもともとは高価で，作るのに不便であったが，それらを生産する安い，容易な方法の発見は，世界第二次大戦が引き起こした新しい性質への興味と一致して，これらの多芸なポリマーの新たな用途への研究のなだれ現象をスタートさせ，それは今日も変わらないでいる。

[Ex. 65] 薬物移送のためのインプラントポリマー

1. 生分解性ポリマーは環境にいいと，我々はみんな聞いたことがある。しかし，それらはがん患者にもいいかもしれない。人体の中でゆっくりと分解し，その過程で抗がん薬を放出するようなポリマーインプラントをデザインするための努力が進行中である。

2. このようなインプラントはいくつかの特別な性質を必要とする。それは外側表面から内に向かって，ゆっくりと分解しなければならない。その結果，インプラント中すみずみまでいきわたっている薬が長時間制御された形で放出されるであろう。ポリマーは全体として水をはじいて，その中の薬を，インプラントそのものの内部と同様に，早すぎる溶解から保護しなければならない。しかし，モノマー―すなわち，ポリマーを形作っているビルディングブロック―間の結合は，水の影響を受けやすくなっていて，ゆっくりと崩壊するべきである。

3. 酸無水物結合―これは，カルボン酸を含む2つの分子が互いに結合して1分子になる時に生成し，この過程で水分子を生じて放出する―が有望な候補である。というのは，水分子がこの無水物結合を，それらを作った逆の過程で，すばやく分解させ，しかも，内部ではポリマー分子がなお撥水性であり得るからである。成分の比を変えることで，1週間から数年もつような表面

侵食ポリマーが合成されている。

4. これらのポリマーディスクは，いま，脳がんの手術後治療として実験的に利用されている。外科医は，脳腫瘍を取り除いた同じ手術で，ポリ酸無水物のいくつかのディスク（各およそ25セント銀貨のサイズ）を移植する。ディスクは，ニトロソウレア（ニトロソ尿素）と呼ばれる細胞を殺す強い薬を含む。ニトロソウレアは普通静脈内に投与されるが，それはどんな場合にも毒性であり，この方法は通常，がん細胞を殺す一方で，体内の他の器官も損傷する。しかし，薬をポリマー中に入れると，薬を身体から，身体を薬から保護する。ニトロソウレアは，およそポリマーの持続時間——この場合，ほぼ1か月——持続する。そして，この侵食されていくディスクは，薬を，がん細胞の潜む直ぐ周囲にだけ運ぶ。薬物移送のポリマー分解法は，食物・薬物局（米国）の承認に向けてよい進展をしつつある。

[Ex. 66]　生体高分子 対 合成高分子

1. 世界における生体高分子の量は，合成高分子のそれをはるかに凌駕している。生物の高分子は，DNA，RNA，タンパク質，炭水化物，そして脂質を含んでいる。（訳注：脂質は，通常高分子には含めない。）DNAとRNAは情報高分子（生物学的情報を暗号化する）であり，一方，球状タンパク質，いくつかのRNA，そして炭水化物は化学的機能と構造用の目的に役立っている。それと対照的に，多くの合成高分子，およびコラーゲン（腱や骨を成す）やケラチン（毛，つめ，羽を成す）のような繊維状タンパク質は，情報あるいは化学機能用というよりはむしろ構造用である。構造材料は，その機械的強度，硬度，あるいは分子サイズなど，分子量，分布，およびモノマーのタイプに依存する性質の故に有用である。

2. 対照的に，情報分子はその主な性質を，単にそのサイズからではなく，情報と機能を暗号化する能力から，引き出している。それらは，異なるモノマーの特別な配列を持った鎖である。モノマーは，DNAではデオキシリボ核酸塩基である；RNAではリボ核酸塩基；タンパク質では，アミノ酸；炭水化物あるいは多糖類では糖である。生体高分子における範例は，鎖に沿ったモノマーの配列が情報を暗号化し，その情報が分子の構造あるいはコンフォメーション（立体配座）を制御し，そしてその構造が機能を暗号化している，ということである。情報高分子はネックレスのようであり，モノマーはそのビーズのようである。

3. RNAとDNAでは，4つの異なるモノマー（違う色のビーズ）がある。情報は，ビーズの色の配列の中に暗号化され，それが，引き継いでタンパク質のアミノ酸の配列を制御する。20の違った色のビーズがある。球状タンパク質は，アミノ酸の配列に依存して，ある特定のコンパクト（密）な構造に折りたたむ。この丸められた形，すなわち構造がタンパク質の機能を決めるものである。線状構造の折りたたみが三次元の形を作り出し，これが形の選択を通じてタンパク質の機能を制御する。

[Ex. 67]　フィルム，膜，塗料

1. ポリマーは，主な機能がある領域から他の領域への小分子あるいはイオンの移動を制御する多くの用途で利用される。例としては次のものがある。酸素を外に，二酸化炭素や水を内に保つ壁を持った容器；基質を，水，酸素，塩から守る塗料；食品を汚染，酸化または脱水から守る包装フィルム；酸素と二酸化炭素の移動をバランスすることで，野菜に呼吸をさせて，長時間の貯蔵や輸送の間新鮮さを維持させるようないわゆる「スマートパック（賢い包装）」；薬物，肥料，除草剤などの制御移送のためのフィルムなど；そして，流体混合物の分離のための超薄膜。

2. これらの多様な機能が達成できるのは，部分的には，溶解—拡散機構による小分子の透過性がポリマーの分子および物理的構造をいじることによって広範囲に変えることができるからである。気体の透過性が最も低いと知られるポリマーは，完全に乾燥したポリビニルアルコールである，一方，最近発見されたポリ(トリメチルシリルプロピン)は今日まで知られている最も高い透過性のポリマーである。酸素ガス(の透過)に対するこれらの極限の範囲は 10^{10} 倍もある。

3. 自由体積，分子間力，鎖の硬さ，運動性などを含む，さまざまな因子が一緒に作用して，この膨大な範囲の輸送挙動を引きおこす。最近の実験研究は大量の洞察を提供しているが，一方，分子力学を用いた拡散過程をシミュレート(模擬計算)する試みは初期段階にある。明らかに，各々の応用に対してポリマーの分子設計におけるガイダンス(指針)の必要性がある。加えて，加工における技術革新も必要である。

[Ex. 68] 室温イオン液体

1. 室温イオン液体(RTIL)は，既存の技術や工程を再評価し，最適化するための多くの機会を提供する。RTIL は，周囲の状況で融解した(典型的には有機の)塩として分類される。これらを普通の有機溶媒と区別する RTIL の固有の性質は，不揮発性，熱安定性，そして調節可能な化学的性質である。これらの性質は，化学工学的応用，とくに気体分離や気体溶解度の促進という点で，RTIL に重要な利点を与えることができる。

2. RTIL は，CO_2 捕獲を目的とする新しい材料(溶媒，高分子，ゲル)の開発に，非常に多才で調節可能な舞台を提供する。CO_2 の溶解度は普通の溶媒に比べて RTIL 中の方が高い。CO_2 と可逆的に結合できるアミンのような特別な錯化剤の添加は，CO_2 の溶解度を 10 倍ほど増加させる可能性を与える。イミダゾリウム由来の RTIL の調節可能な化学的性質は，また，新しいタイプの CO_2 選択性高分子膜の創製を可能にしている。

3. CO_2 選択の性質が与えられると，RTIL を膜の選択性の成分として用いることに大きな興味が出てきた。膜構造の中に RTIL を用いる直接的な方法は，イオン液体の担持された膜を採用することである。一般に，担持液体膜は，高分子あるいは無機の担持体に固定化された液体から成る。それらは一般的に，従来の高分子膜よりも大きな気体透過性を与える。というのは，密な液体膜を通しての気体の拡散は，ゴム状あるいはガラス状高分子を通した場合よりも，しばしばはるかに速いからである。

4. しかしながら，従来の担持液体膜は，気流中への液相の蒸発による損失，その結果膜本体と性能の低下をもたらすという問題がある。担持 RTIL 膜は，この制限を免れる：RTIL 成分は蒸発しないからである。いくつかのさまざまなイミダゾリウム由来の RTIL の評価によって，担持 RTIL は従来の高分子膜よりも勝る透過性と CO_2/N_2 選択性を有するということが明らかになった。

[Ex. 69] ブロック共重合体ミセル

1. ブロック共重合体が選択性溶媒中で自己集合してミセル状凝集体になる能力は，さまざまな科学技術，例えば医薬移送と治療，触媒，分離，化粧品，および食品科学において，潜在的な関心がよせられている。多くの点で脂質，界面活性剤，および小分子両親媒性種と類似しているけれども，ブロック共重合体は性質の調整可能な点で多くの利点がある。

2. ある与えられた AB ジブロック共重合体で，A の良溶媒中では，標準的なミセルの形は，球，虫状，ベシクル(小胞体)である。平衡条件下で形状の選択を支配する因子は，かなりよく知

られている。例えば，Bと溶媒間の界面張力を増加すると，より大きなミセルおよびより平たい界面への変化を促すが，一方，十分に溶媒和されているAのコロナブロックの密集は，反対の方向に働く。

[Ex. 70] セレンディピティ

　セレンディピティの恩恵を受けるのに備えることができる1つの道は，その選ばれた研究分野での注意深くそして集中した勉強(研究)による。アメリカの物理学者ジョセフ・ヘンリーはパスツールの格言を意訳して，次のように言った，「大きな発見の種は私たちの周りにいつも浮遊しているが，それらを受け入れるのに十分準備のできている心(人)にのみ根をおろす」と。例えば，フレミングは彼のペトリ皿に胞子が浮かんだ時点で，抗菌剤を探していたわけではなかったけれども，彼は微生物学に通じ(多くの本を読み)，熟練していて(訓練を積んでいて)，(その結果)バクテリア培養液の中に透明な領域が偶然のかびの移植でできたという意味を容易に認識できたのである。

[Ex. 71] ポリエチレンとポリプロピレンにおけるセレンディピティ

　1. ポリプロピレンおよびポリエチレンは巨大な分子であるばかりでなく，巨大な量で，私たちの世界で生産され，使用されている。世界で毎年生産されるポリエチレンの量は，数千億キログラム(数億トン)と見積もられ，多くの応用に利用されている。Morawetz(モラベッツ)はポリエチレンの発見を面白く詳しく述べている。この発見は，英国のICI社(Imperial Chemical Industries Ltd.)で第二次世界大戦前の数年，高圧下での化学反応を開発しようとする決断の結果として生まれた。ICI社はこれを基礎研究への介入とみていた，そして興味あることに，当時1920年代後半にDuPont(デュポン)も同じような決断をして，Wallace Carothers(ウォレス・カローザス)を雇って，やはり商業製品を目指すことなく，(基礎)研究を行った。DuPontの決断はナイロンの発明に導き，一方ICI社の決断はポリエチレンへと導いた。基礎研究を行うのは，悪い考えではないのだ！

　2. エチレンの重合は，ICI社で高圧下ベンズアルデヒドにエチレンを付加させるという誤った試みの中で偶然起こった。ベンズアルデヒドは変化なく回収されたが，ワックス状の固体が得られ，これはエチレンの比較的低分子量のポリマーであることがわかった。結果を改良しようとする何回かの実験のあと　－そして，爆発のためにいくらか遅れたが－，高分子量のポリエチレン，(すなわち)最初に得られたワックス状の物質とは対照的な固体物質，が高圧のエチレンと小濃度の酸素を用いて作れるということが明らかになった。酸素についての洞察はもう1つの過ちから生じた。－(すなわち)装置の漏れで，不注意で空気が入ったのである。

　3. われわれは読者に，ずさんなやり方で行った実験や，不可能な目標を目指すことが化学での成功に必要であると信じるように，間違った方向に導くつもりはない。しかし実際，Ziegler(チーグラー)とNatta(ナッタ)にノーベル賞(訳注：1963年)をもたらした研究のところで再び偶然の発見を見ることができる。これらのセレンディップな(偶然の恩恵をうける)発見が可能となったのは，ただ，これらの化学者が革新家そのものであり，専門に熟達していて，彼らの実験結果の驚くべき意味を十分理解するだけの準備ができていたからである。(人は)Louis Pasteur(ルイ・パスツール)の有名な引用句を思い出させられる。"観察の分野では，チャンスは準備のできている心(人)だけを助ける"と。

　4. プロピレンの立体特異性重合も，実際にはエチレンを重合させる新しい方法の偶然の発見

で始まる。これは，Kahl Ziegler（カール・チーグラー），触媒が専門の著名なドイツの教授，の実験室にあった化学装置の汚れから生まれた。Ziegler 実験室の一連の実験は，世界第二次大戦のすぐ後で始まり，ICI の研究で用いられたのとは別の方法でエチレンを重合させようとした。実験に使ったオートクレーブ（高圧釜）の 1 つがブテンを高収率で生産するのにとくに効果的であることが発見された。これを追跡すると，最終的に，ステンレスのオートクレーブの洗浄で生成したコロイド状のニッケルの存在とその後で実験に用いられていたリチウムに触れていたことにたどりついた。

5. 結果は，周期表の金属の全域にわたる探索を刺激し，この結果を最適化するような他の金属を見つけようとした。科学研究の皮肉の 1 つとなるが，多くの金属のテストはエチレンからブテンの生産という目的を達成するには不成功であったが，むしろエチレンを重合させるという本来の目的を達成した点で結果としては成功であった。ある金属，ジルコニウムとチタン，の添加がエチレンを高分子量ポリエチレンに変えたのである。そして直ちにわかったのは，この新たな方法で生産されたポリエチレンは ICI 高圧法で生産されたものと違っていることであった。

6. 赤外分光分析の早い段階での利用によって，新しいポリマーは ICI 法で生産されたものよりも少ないペンダント（垂れ下がり＝側鎖分岐）メチル基[*]を持っているように思われた，そしてまた，それはより高い温度で軟化した。したがって，Ziegler 実験室の研究から，ある触媒が開発されたことになる：すなわち，フリーラジカルを用いた開始段階で生成した普通のポリエチレン（訳注：低密度ポリエチレン LDPE）とは重要な点で違っているポリエチレン（訳注：高密度・高結晶性ポリエチレン HDPE）を生成させるような触媒である。

[*]訳注：実際の分岐はエチルまたはブチル基が多いが，その末端のメチル基が IR で検出される。

INDEX

演習問題［Ex.］の［注］にあげた大部分の英単語・句とその対訳をアルファベット順に並べた。
数字は，対応する［Ex.］の所在ページを表す。

A

abandon 捨てる 42
abbreviate 略する 77
abhorrent 相容れない 73
able to できる 95
absolute zero 絶対零度 67
absorb 吸収する 15, 110
absorbance 吸光度 79
absorbtion 吸収 18, 110
abstinence 禁欲 103
abundant 豊富な 35, 108
accept 受ける 91
accompany 伴う 104
accomplish 成し遂げる 48
according to によれば 64, 99, 100
account for を説明する 83
accounts 記事 34
accuracy 正確度 28
achievement 偉業，達成 100
acidity constant 酸性度定数 91
acidosis アシドーシス（酸性血症） 90
acquire 獲得する 62
addictive 習慣性の 103
addition polymers 付加重合体 113
additive 添加物 105
address 処置する 105
adhere to に固執する 20
administrate 投与する 101
admit 容れる 88
adrenocortical 副腎皮質の 110
advantage 利点 81
affair 事柄 21
after-image 残像 23
aftertaste あと味, 97
against に反する（逆らった） 72
aggregate 凝集体 120
alchemy 錬金術 61
alert 油断のない，機敏な 70
alertness 警戒心 102
alien 相容れない，外国の，異なる 73
align 並べる 56, 63
alleviate 軽減する 75
Allied 連合国の 103

allow to させる 47, 75, 122
alloy 合金 20, 82
allyl-resonance アリル共鳴 114
alternately 交互に 80
alternating 交互の 68, 115
alternative 代替品 82, 104
aluminosilicate アルミノ珪酸塩 88
amazing 驚くべき 48
ambient 周囲の 118
ameliorate 良くする 84
among others いろいろある中で 53
amphiphile 両親媒種 76, 120
amphiphilic 両親媒性の 76
amphoteric 両性の 76, 107
amplitude 振幅 77
amu 原子質量単位 50
analgesic 鎮痛剤 95
ancestry 祖先 97
and so forth など 81
and such その他同様なもの（人） 96
anesthesia 麻酔 66
anhydride （酸）無水物 116
anion アニオン（陰イオン） 68
annihilation 絶滅 65
antacid 制酸剤 84
anthroscopy 内視鏡を用いる診断 82
anti-inflammatory 炎症抑制剤 95
antibacterial agent 抗菌剤 120
antiperspirant 発汗抑制剤 69
antipyretic 解熱剤 95
antiseptic 防腐剤 109
apparatus 装置 122
apparent 見かけの 99
apparently 見かけ上，と思われる 98
appear に見える，思われる 72
appliance 器具 36
apt to しがちである 32
archaeological 考古学的 65
armed with で武装した 22
arrangement 配置 53
array 並び 114
arthritis 関節炎 96
artificial 人工の 97

as 〜 as　と同じぐらい〜　36, 88
as early as　ほども早くに　61
as if　あたかも…のように　80
as soon as　や否や　72
as with　と同様に　114
assembly　集合　48
asset　資産　95
associate　同僚　98
associated with　に関連した　19, 107
astoishing　驚くべき　48
astronomer　天文学　15
astrophysicist　天体物理学者　18
at one's disposal　の自由に（使える）　15
atmosphere　大気　15
atmospheric　大気の　71
attach　結合する　46, 56
attached with　を付した　68
attain　到達する　29
attractive　魅力的な　54
attributable to　に帰すことができる　31
availability　入手可能であること　82
avalanche　なだれ, 殺到　115
average　平均して〜である　50

B

baking powder　ふくらし粉　67
baking soda　重曹　67
bark　（木）皮　49, 95
basal metabolic rate　基礎代謝率　89
be compared to　と同じぐらいである　45
be composed of　から成る　18, 20
be interested in　に興味がある　20
be responsible for　の原因となる　53
bend　曲がり　51
bending　変角の　81
beneficial　有利な　85
benefiit　恩恵を受ける　57, 120
besides　のほかに, に加えて　79
bestow　授ける　62
biodegradable　生分解性の　116
biological　生物の　52
bioluminescence　生物発光　25
biomass　バイオマス（生物資源）　42
biopolymer　生体高分子　117
biorefinery　生物資源変換　104
blame　非難する　34
blight　暗い影　36
block copolymer　ブロック共重合体　120
bombard　衝撃（爆撃）する　64
bone-dry　からからに乾燥した　118
borne in mind　心に留める　32

brass　真ちゅう　20
break with　断絶する　22
breath　息　66
breathe　呼吸する　66
bugbear　怖いもの, お化け　72
build on　に頼る　61
building block　構成単位　46
bulk　本体, 内部　76
bump into　ぶつかる　47
by a factor of　〜倍で　119

C

canonical　標準の　120
captive　捕獲（包接）した　87
capture　捕獲　119
carbonated beverage　炭酸飲料　67
carcinogenicity　発がん性　97
cation　カチオン（陽イオン）　68
cause　させる　45, 55, 70, 80
cavity　穴　88
ceiling temperature　天井温度　113
celestial　天の　18
cellular phone　携帯電話　77
central nervous system　中枢神経系　102
certain　ある　65, 80, 81
chair conformation　いす形配座　111
channel　溝, 水路　88
characterization　特性化　76
characterize　特性化する　77
charm　（腕輪の）飾り　115
chemical　化学薬品　104
chicken pox　水ぼうそう　96
chirality　キラリティ（掌性）　40
chromosome　染色体　44
chronological　年代順の　65
circumvent　免れる　109, 119
classification　分類　106
classify into　に分類する　107
clathrate　クラスレート（包接体）　85
cleanser　クレンザー（洗剤）　109
clergyman　牧師　95
closed　括弧内にに入れた　91
clotting　凝固　89
cluster　かたまり　46, 69
cluster　群がる　109
coagulate　凝集する　30, 50
code　暗号にする　44
coefficient　係数　69
cohesive energy　凝集エネルギー　76
coin　造る　49
coincide with　と一致する　115

colon 結腸 96
combination 組合せ，結合 32, 52, 68
combustible 可燃性の 32
combustion 燃焼 34, 69
common logarithm 常用対数 92
common 共通の，普通の，常用の 29, 79, 92
comparable 比較しうる，同等の 31, 83
compared to に同じようである 114
compared with と比較して 76
compatible with と両立できる 106
compete against と対抗する 52
competitor 競合者 101
complementary 相補的な，相補う 26
complex 複合体，錯体 47
complex 複雑な 20, 42, 47
complexing agent 錯化剤 119
complication 併発症 96
component 成分 74
compose 構成する 68
composed of から成る 43
composition 組成 15, 32
compound 化合物 63, 68
comprehensive 包括的な 84
compression 圧縮比 37
comprise 構成する 88
concentrated 濃い 30
concentration 濃度 32, 79, 91
concept 概念 28
conception 概念 70
concerning に関して 32
condition 条件 74
confirmed 頑固な 22
conjugate 共役の 91
conjugated 共役した 50, 78
connectivity 接続性 78
consequence （必然の）結果 114, 121
conservation 保存 72
consist of から成る 21, 43, 63, 64, 68, 99
constipative 便秘剤 85
constituent 構成成分 39, 74
constituent 成分の，構成する 43
constitute 構成する 18
container 容器 68
contamination 汚れ，汚染 118, 122
contemporary 同時代の，現代の 82
contract 縮める 80
contradictory 矛盾した 115
contribute 寄与する 61
convenience 便利，都合 74
conventional 伝統的な，従来の 119

conversion 変換 19
conversion factor 変換因子 69
convert into （to）に変換する 43, 64
convert 変換する 19, 43
convince 納得させる 73
coordinate 同格の 72
correspond to に相当する，対応する 70
corresponding に対応する 80, 107
corrode 腐食する 35
corrosive 腐食性の 34
cosmology 宇宙論 73
counterbalance 相殺する 88
covalent 共有結合の 68
covalent bond 共有結合 92
creation 創造 94
cretinism クレチン病 89
criterion 規準 71
critical 決定的の 45
critically 決定的に 75
cross coupling クロスカップリング 57
cross section 断面 75
cross-link 架橋する 51
crucial 決定的な，致命的な 45, 87
crude 粗い，生の 28, 49
crumble ぼろぼろになる 36
crust 殻 87
crystallization 結晶化 56
curd 凝乳（カード） 109
curdy 凝乳状の 108
curiosity 好奇心 85

D

dangerous 危険な 32
dangle ぶら下がる 115
date 日付をする，年代を推定する 65
dating 年代を算定する 66
debatable 異論のある 36
decay 崩壊する 64, 66
decimal 十進の 63
decode 暗号を解く，解読する 112
deficiency 不足 53, 89
definite 一定の 19
definition 定義 29, 91
degradation 退化，分解，低下 72, 119
degrade 分解する 116
dehydration 脱水 118
demagnetize 消磁する 30
depend on に依存する 79, 97
dependent 依存性の 86
depending on に依存して 100, 107
deplete 枯渇させる 90

depletion 枯渇 69
deposit 堆積床，鉱床 85
derive from から引き出す 69
descriptive 記述的な 63
destructive 有害の 34
detergent 洗剤 69
detonate 爆発する 37
detonation 爆発 65
develop 発現する 30
diagnose 診断する 65
diagnostic 診断の 78, 81
dialysis 透析 110
dictim 格言 120
differentials 微分 74
differentiate 区別する 118
diffract 回折する 78
diffraction 回折 18
diffuse 拡散する 30
diffusion 拡散 75
difunctional 2官能性の 106
dig out 調べ上げる 59
digest 消化する 66, 110
digestion 消化 110
dilute 薄い 30
dimer 2量体（ダイマー） 37
dimerization 2量化 37
dipolar 双極性の 107
directly proportional 正比例の 67
disaccharide 2糖 43, 111
disarm 軍備を解く，軍備縮小する 65
disarray 乱雑 56
discard 捨てる 36
discrete 不連続の 77, 83
disorderly 無秩序な 114
disperse 分散させる 105
dispersion 分散（液） 50
dispersion force 分散力 87
dissipation 散逸 72
dissolve 溶ける 91
distinguish 区別する 20
distinguished 著名な 122
diverse 多様な 118
diversity 多様性 99
divide into に分類する 68, 76
dogma 教義，定説 112
dominance 優勢 35
donate 与える 91
dough パン生地 67
downright まったくの 26
drawing 延伸 56
ductile 延性の 35

ductility 延性 83
duplication 複製 47
duration 持続期間 116
dwindle away 減少してなくなる 65

E

each other お互い 26
E-coli 大腸菌 48
effuse 流出する 30
either A or B AかBかどちらか 43, 52, 55, 101, 107
elasticity 弾性 114
elastomer 弾性体（ゴム） 114
electrical conductivity 電気伝導度 87
electromagnetic radiation 電磁線 64, 76
electronegative 電気陰性の 53
electronegativity 電気陰性度 91
eliminate 取り除く，脱離する 34, 105
elongation 伸び 54
emergence 出現 98
emission 放射 18
emit 発する 15, 64
emphasize 強調する 61
enantiomer 鏡像体 40
encapsulate 要約する 59
encode 符号（暗号）化する 117
encompass 包含する 21
encourage 促す 56
endoscope 内視鏡 82
enforce 強める 56
enough that に十分な 35, 98
ensuing 引き続く 82
enteric-coated 腸溶性コートされた 96
entrench 立場を守る，強固にする 105
environment 環境 69
eon 地質時代の最大区分，永劫 85
equal に等しい 63
equilibria （複数）平衡 71
equilibrium （単数）平衡 29, 91
escape 逃れる 21
essential 必須の 89
essential oil 精油 39, 98
establish 設立する 97
ethical 倫理的な 65
evacuate 空にする，脱気する 30
even 偶数の 108
exceedingly 非常に，過剰に 47, 87
exceptionally 例外的に，異常に 56
exclusive right 占有権 102
excrete 排泄する 98
exemplify 例証する 61

exhale （息を）吐き出す　66
existing　既存の　118
exothermic　発熱の　113
expedition　旅行　24
expel　放出する　37
exploitation　利用　27
exploration　探索　122
explore　探求する　61
explosion　爆発　121
explosive limit　爆発限界　32
extend　拡大する　15, 75
extinguish a fire　火を消す　67
extract　抽出物　95, 98, 102
extraction　採取，抽出　71, 82, 107
extrapolate　外挿する　68
extrude　押し出す　55

F

fabulous　伝説的な　22
fall　崩壊する　116
fall short of　まで届かない　59
far from　にはほど遠い　45
far-reaching　遠くに及ぶ，広範囲の　73
feasibility　（実行の）可能性　71
feedstock　原料　104
feldspar　長石　88
fermentation　発酵　101
fertilizer　肥料　118
fission　（核）分裂　64
flash point　引火点　32
flatulence　腹の張り　84
flourish　栄える　61
fluolescent　蛍光性の　103
foaming agent　発泡剤　115
followed by　に引き継がれる　61
foray　侵略　121
force to　（強制的に）させる　64, 114
foresee　予見する　98
foreshadow　予示する　73
former, latter　前者，後者　31
formulate　定式化する　73
formulation　処方　85
fossil　化石　65
fossil fuel　化石燃料　27
foundational　基本の　63
framework　骨組　85
frequency　周波数　77
functional group　官能基　52, 94
fundamental　基本的な　61
furthermore　さらに　63
fusion　融合　64

G

gastric juice　胃液　111
gastric ulcer　胃潰瘍　96
gastrointestinal bleeding　胃腸の出血　96
generalization　一般化，一般則　21, 73
generate　発生する　64
generic drug manufacturer　ジェネリック薬品製造会社　102
genetic　遺伝の　41
genetically engineered　遺伝子工学の　101
genuine　真の　65
geology　地質学　70
geometry　幾何学，幾何異性　108
give off　発する　32
gland　リンパ腺　89
global warming　地球温暖化　67
globular　球状　117
go out　消える　67
goiter　甲状腺腫　89
gradient　勾配　30
grain boundary　粒界　83
granite　花崗岩　88, 115
great-great-grandfather　曽曽祖父　97
greenhouse　温室　70, 86
group into　分類する　63
growing chain　成長している鎖　50

H

half-life　半減期　66
half time　半減期　75
halo　かさ，後光　23
harness　利用する　62
harvest　収穫する　49
head　頭となる，率いる　72
head-to-tail　頭-尾　99
heat capacity　熱容量　73
help　させる　69
herbicide　除草剤　118
hire　雇う　121
hodgepodge　ごた混ぜ　75
homogeneous　均一な　20, 67, 74
how far　どこまで　71
human being　人間　20
humdrum　平凡な　22
hydrate　水和物　85
hydrogen bond　水素結合　53
hydrophilic　親水性の　76, 109
hydrophobic　疎水性の　76, 109

I

identical　同じ　94

identifiable 同定できる 69
identify 同定する 15, 68, 78, 95
if —, then~ もし—なら, ~ 27
ignore 無視する 62
immediately 直ちに 96, 114
immense 膨大な 100
immobilize 固定化する 119
impact 影響 27, 62, 65
impart 付与する 118
impediment 障害 114
implant インプラント（する） 116
implement 実行する 112
implication 意味, 連携 65
imply 意味する 70, 106
improvement 改善 21
in accord with と一致して 72
in accordance with と一致して 74
in addition 加えて 107
in common 共通に 88
in comparison with と比較して 36
in contact 接触している 29
in contrast 対照的に 71, 117
in order to ~するため（よう）に 69
in order of の順に 63
in practice 実際上 71
in that という点で 83
in time やがて, 間に合って 28,
in turn 引き続いて 34, 112, 117
inadvertently 不注意で 121
incorporate 取り入れる 27
incredulous 懐疑的な 73
independently of に無関係に 74
indiscriminately 無差別に 116
individual 個々の 46
inert 不活性な 63
infect 感染させる 41, 103
inflammable = flammable 可燃性の 32
informational 情報の 117
infrastructure 構造基盤 104
ingestion 摂取 84
ingredient 成分 95
inhabit に住む 24
initiate 始める 26
initiation 開始 122
innovation 革新 118
innovator 革新家 121
inprinting 刷り込み 23
insight 洞察 121
instruction 指示 44
insulation 絶縁 115
insuperable 克服できない 74

integral 欠くことのできない 73
integrity 完全, 本来の形 90, 119
interaction 相互作用 15, 85
interconversion 相互変換 74
interfacial tension 界面張力 120
intermediary 仲介者（物） 42
intermediate 中間の 74
intermolecular 分子間の 118
interpret 説明・解釈する 37, 81
interpretation 解釈 81
interstellar 星間の 18
intestinal 腸の 110
intractable 扱いにくい 104
intramolecular 分子内の 107
intravenously 静脈内で 116
intrinsic 固有の 118
intuition 直感 21, 72
invaluable 貴重な 34
invasion 侵略 103
invent 発明する 70
invisible 目に見えない 67
ionic liquid イオン液体 118
ionizing power イオン化力 64
irradiate 照射する 79
irritation 刺激 95
isolate 単離する 107
it seems と思われる 73
it — that ~: ~は—である（—を強調） 45

K

keep from から防ぐ 51, 115
keep ~ going ~が活動を続ける 69
kinetic energy 運動エネルギー 68, 70
kink ねじれ 51

L

ladder はしご 46
lamina 薄層 75
lattice 格子 78
laxative 便通剤 85
lethargy 無気力 103
likely to ~らしい 108
link A to B AをBに結びつける 64
lipophilic 親油性の 76
literally 文字通りに 45, 70
locality 地方, 産地 35
logical 論理的な 21
long-familiar 長く親しんできた 73
longevity 長寿命 104
lurk 潜む 116

M

macromolecule 巨大分子，高分子 117
macroscopic 巨視的 61, 64
made of からできている 62
made up of から成る 43
magnification 倍率 46
make up を成す 89, 117
malleability 展性 83
malleable 展性の 35, 82
manipulation 処理，操作 82, 118
mass-energy 質量―エネルギー 18
matter もの，物質 18
measles はしか 41
measurable 測定可能な 28
megaton メガトン 86
membrane 膜 119
menopause 閉経 90
mentor 助言者 59
messenger 伝令 48
metabolic 代謝の 71
metabolism 代謝 82
metabolite 代謝物 39, 71
metabolize 代謝する 98
metallurgical 冶金の 71
metaphysical 形而上学の 23
microscopic 微視的な 64
migration 移動 118
millennium 一千年 65
mine 鉱山，採掘する 22, 35
minimize 最小化する 27
minimum 最低 68
minute 微小の 103
mix-up 混乱 26
modification 変形，修飾 63
modified 修飾された 41
molar absorbtivity モル吸光係数 80
mold カビ 39
mold 鋳型 47, 115
mold 成形する 114
molecule-specific 分子特有の 80
monosaccharide 単糖 43
motionless 動かない 68
mount 登る，始める 88

N

naming 名前を付ける（命名の） 68
nausea 吐き気 103
necessarily 必然的に 28
necessary to に必要な 79
nestle （巣を作るように）擦り寄る 51
nonvolatility 不揮発性 118
not always いつも…ではない 102
not only ～ but (also) ― (as well) ～だけでなく，―も 77, 92, 104
not to be outdone 負けることない 22
not ―, but (rather) ～：― でなくて，(むしろ)～ 20, 72
nuclear power units 原子力設備 83
nuclei （複），nucleus（単）核（原子核） 64
numerical 数の 69

O

object 物体 20, 29
objective 目的 71
odd 風変わりな 22
of の内 89
oligomerize オリゴマー化する 105
once ～ 一旦～すると 30, 71
one, the other 一方，他方 29, 32, 98
one another お互いに 87
one with the other 互いに 56
one…another あれやこれやの 23
ongoing 進行中の 98
open-chain 開鎖の 99, 111
operational 操作の，使用可能な 29
opportunity 機会 56
optimize 最適化する 105, 118
or すなわち 32
orbit 回る 62
ordering 秩序化の 113
ore 鉱石 35, 82
organize 組織する，整理する 63
orientation 配向 56
originate 生じる 34
orthodoxy 伝統 22
osteoporosis 骨粗しょう症 90
other than と別の 122
over-the-counter medication 店頭市販薬 84
overwhelming 圧倒的な 35, 72

P

pancreatic 膵臓の 111
paradigm 範例 117
parallel to に平行な 75
paraphrase 意訳する 121
participate in に関与する 53
particular 特定の 15, 36
patch パッチ（布切れ） 115
path-length 光路長 80
pendant ペンダント（垂れ下がった） 122

penetrating power 浸透（貫通）力　64
perchloric acid 過塩素酸　93
percolate しみ出る　85
performance 性能　105
perfume 香料をつける　109
periodic 周期的な　63, 78
periodic law 周期律　63
permeability 透過性　118, 119
permit させる　72
perpendicular 垂直の，直交する　75
persist 固執する，持続する　42
perspective 見通し　61, 70
petri dish ペトリ皿（シャーレ）　56, 120
pharmaceuticals 薬品　84
phenomena, phenomenon 現象　61
phosphate リン酸エステル，リン酸塩　46
phospholipid リン脂質　90
photosynthesis 光合成　42, 66
physiological 生理学の　28, 102
platform 演台，舞台　119
plow 鋤（すき）　82
polarity 極性　91
polycondensation 重縮合　52
polymerase ポリメラーゼ　47, 106
polysaccharide 多糖　43, 111
postoperative treatment 手術後治療　116
potent 有力な　86
poverty 貧乏　22
ppb = part per billion 10億分の1　87
precipitate 沈殿　102
precision 精度　28
preclinical 前臨床の　101
precursor 前駆体　99
predecessor 先行者　61
predict 予測する　63
prefix 接頭語　63
pregnancy 妊娠　96
prejudice 先入観，偏見　72
prematurely 早すぎて　116
premise 前提とする　88
prepared 準備ができている　121
preposterous 途方のない　26
present : 存在する　15, 68
preserve 保存する　72, 112
pressure gauge 圧力計　68
prevalent 広く行き渡っている　66
prevent from から防ぐ　51
primarily 主に　20, 69
primitive 初期の　118
prior to の前に　102
probability 確率　73

processing 加工　118
produce 生産する，生じる　107
productivity 生産性　34
product 生成物　69, 91
promote 昇位させる　79
prone to しやすい　37
pronounced 著しい　102
propagate 伝播する　32, 61
proposal 提案　63
propose 提案する　63
proprietary 専売の　85
prototype 原型　20
provided that の条件で　68
puckered ひだになった　111
pumice 軽石　109
punctuate 中断する　24

Q

qualitative 定性的　84
quanta, quantum 量子　73, 77
quantify 定量する　15
quantitative 定量的な　30, 84
quantize 量子化する　73, 80
quarter 1/4, 25セント銀貨　116
quip 皮肉る　26

R

radiant energy 放射エネルギー　18
radioactivity 放射能　64
ranch 農場　24
ratrio of A to B A対B比（A/B）　65
reactant 反応物　69
rearrange 転移する　94
reassess 再評価する　118
rebellious 反体制の　22
reciprocal 逆数　80
recognition 認識　72
recognize 認識する　26
reconcile 和解させる，調和させる　73
recount 詳説する　121
recur 繰り返す　63
refer to を指す，言及する　42, 92
reflectivity 反射性　83
regardless of に関わらず　99, 106
related to に関係がある　91
relativity 相対論，関連性　19, 71
relax 緩む　114
release 開放する，放出する　51, 116
relevance 関係　70
relieve 軽減する　53
remain 残る，持続する　70

renewable 再生可能な 104
repeating unit 反復（繰り返し）単位 113
repel 反発する，はじく 116
repetitive 繰り返しの 56
replace 置き換える 88
replacement 代替品 54
replacement 置換 88
replicate 複製する 45
replication 複写，複製 103, 112
represent 表す 68
resolve 決定する 34
resonance 共鳴 28
resonance-stabilized 共鳴安定化した 38
respectively それぞれ 61, 114
respiration 呼吸 66
respiratory distress 呼吸困難 86
respire 呼吸する 118
response 応答 28
responsibility 責任 62
responsible for の原因となる 31, 96
rest on に頼る 94
restore 取り戻す 34
result in となる，を生じる 64, 88, 119
resulting 生じた 72
retina 網膜 23
revert 戻す 51
revolution 革命 61
revolutions 回転数 48
Reye's syndrome ライ症候群 96
ribosome リボソーム 48
rigid 硬い 68
rigorously 厳密に 20
ring-opening 開環の 52
rinse すすぐ 109
routine おきまりの 21
rub 擦る 49
rung （はしごの）段 46
rust 錆びる 36

S

safe 金庫 22
saliva 唾液 111
salix やなぎ属 95
saponification けん化 108
saturated 飽和の 86, 108
scattering 散乱 78
scour 磨く 109
scum かす 109
seafloor 海底 85
searching 厳重な 73
secretive 秘密的な 61

sediment 堆積物 85
seemingly 見かけは 115
seize 捕らえる，差し押さえる 95
selective 選択的 119, 120
self-assemble 自己集合する 120
self-duplication 自己複製 40
semiconductor 半導体 87
sensation 感覚 28
sequence 順序，連鎖 48, 112, 117
sequestration 隔離 86
serendipity セレンディピティ 120
sesquiterpene セスキテルピン 99
set on fire 着火する 32
share 分け合う，共有する 26, 69
shipping 輸送 118
SI system 国際単位系（SI系） 63
side effect 副作用 101
significant 意味のある，重要な 119
significantly かなり 102
silicon シリコン（珪素） 115
silicone シリコーン 115
sinter 焼結する 83
six-sided 六辺の 111
skyrocket 急騰する 88
slice 切れ目 49
sloppy ずさんな 121
slow down 遅くなる 68
slugginess 無精 89
small intestine 小腸 96, 111
smelt 精錬する 82
smother 窒息させる 86
so that するように 63
solipsistic 唯我論の 24
solubilize 可溶化する 109
solute 溶質 30
solution 解，解決 71
solution-diffusion 溶解―拡散 118
some ある，いくつかの 85, 86
sophisticated 精巧な，綿密な 57, 82
span generations 何代にも亘る 45
spark 引き起こす 115
specific resistance 比抵抗 87
specific 特有の，特定の 15, 68, 80
spectroscopy 分光学 15
spinnert 紡糸口金 55
splash はね飛ぶ 98
split 分裂する 64
spontaneity 自発性 71
spontaneously 自然に，自発的に 29
spore 胞子 120
spur 拍車をかける 106

stand in the way of 立ちはだかる 114
stand out 抜きん出る 71
stem from から生じる 106
stereospecific 立体特異性の 122
stereotypical 紋切り型の 24
stimulant 刺激剤 102
stimulate 刺激する 66, 81
stoichiometric 化学量論的な 32
stoichiometry 化学量論 87
strand より糸 46
stretch 引っ張る，伸ばす 56, 80, 114
stretching 伸縮の 81
stroke 鼓動，卒中 82
structural feature 構造的特長 94
struggle with 苦闘する 65
struggle 戦う 45
strung (string 糸になる) の過去分詞 112
sublimation 昇華 67
substance 物質 18, 118
subsurface 表面下の 85
succeeding 続く，次の 112
such — that ～: ～ほどの — 32
suffer from を苦しむ 119
sufficient that するに十分な 86
sufficient to に十分の 32
sufficiently 十分に 46
sugar 糖 46
superacid 超強酸（スーパーアシッド） 93
supported 担持（支持）された 119
support 支持体，担持体 119
surface tension 表面張力 76
surface-eroding 表面侵食する 116
suspicion 疑い 97
swallow 飲み込む 111
swelling 腫れ 95

T

tackify 粘着性を増す 115
tacky べとべとした，粘着性の 50
take a brief look at を少し見る 64
take advantage of を利用する 78, 81
take place 起こる 23
tangible 実体的な 65
tangle 絡む 56
template テンプレート（型板） 47
temporary 仮の 25
tendency 傾向 29
tendon 腱 117
tensile strength 引っ張り強さ 54
that is すなわち 28
the Curies キュリー夫妻 64

the incredulous 懐疑派 73
the same ～ as —: —と同じ～ 75
the — （比較級），the ～（比較級）: —であればあるほど，より～ 32
therapeutic 治療の 65
therapeutics 治療（学） 120
thin 細い 56
though （副詞）しかし，けれども 79
throat のど 84
thyroid 甲状腺 89
thyroxin チロキシン 89
tinge 薄い色 109
tolerance 抗薬力，寛容 103
traditional 伝統的な，従来の 119
transcription 転写 47, 112
transform 変化する 69
transformation 変化，変換 30, 105
translation 翻訳 47, 112
transmission 移送 118
transport 輸送 118
transportation 輸送 105
triatomic 3原子の 87
trigger 触発する 86
trillion ［米］兆（10^{12}） 44
troops 軍隊 103
tubular-flow reactor 管流動反応器 75
tunable 調和可能な 118
turn out 判明する 74, 89
twin 双子の，対の 74
twist ねじる 46

U

ultimately 究極的に 65
ultrathin 超薄の 118
unabated 減じない，変わらない 115
unbranched 分岐のない（直鎖の） 108
under way 進行中の 116
underlying 基礎となる 61, 64
understatement 控えめな表現 69
undisciplined 規律にしばられない 22
undoubtedly 疑いなく 28
unflappable 冷静な 22
unorthodox 非正統的な 22
unravel 解決する 61
unsaturated 不飽和の 79, 108
unshakably ゆるぎなく 22
unwind ほどく 47
upgrade 格上げする 104
upper atmosphere 高層大気 66
upset ひっくりかえった 84

V

valence orbital　原子価軌道　92
vehicle　車　105
versatile　多芸多才な　15, 115, 119
verification　証明　19
vertical　縦の　63
vessel　容器　30
viability　生育可能, 成熟　105
vibration　振動　78
virtually　実質的に　15
visualization　可視化　82
vulcanization　加硫　50
vulnerable to　（の影響を）受けやすい　34

W

wag back and forth　前後に振る　80
warning　警告　36
waste disposal　廃棄物処理　65
wavelength　波長　77
wavenumber　波数　80
well-fed　栄養の十分な　48
well-solvated　よく溶媒和された　120
what to do with　をどうするか　106
whence　そこから　72
whereas　一方で　83, 92
while（whereas）　一方で　27, 92, 106, 108
widespread　広汎な　102
willow　柳　95
withdraw　引っこめる　56
withdrawal symtom　禁断症状　103
whether A or B　AでもBでも, AかBかどちらか　20, 43
worm　虫（のような形）　120

Y, Z

yield　生じる, 産する　51, 108
Z-(zusammen)　cis- と同じ意味　50
zwitterion　ツヴィッターイオン　107

著者略歴

伊藤　浩一
　1939年　福井県生まれ
　1966年　名古屋大学大学院工学研究科博士課程修了後，助手，
　　　　　助教授をへて，豊橋技術科学大学教授，豊橋技術科学
　　　　　大学名誉教授　工学博士
　専　門　高分子合成，分子制御化学

蒲池　幹治
　1934年　福岡県生まれ
　1961年　大阪大学大学院理学研究科博士前期課程修了後，東洋
　　　　　レーヨン株式会社，大阪大学理学部高分子科学科助手，
　　　　　助教授，教授をへて，現在，大阪大学名誉教授，前福
　　　　　井工業大学教授　理学博士
　専　門　高分子合成，物理有機化学

化学英語文献への誘い──英語演習を通して化学を学ぶ

2011年3月25日　初版第1刷発行
2024年3月20日　初版第4刷発行

　　　　　　　　　　　　　　　　Ⓒ　著　者　伊　藤　浩　一
　　　　　　　　　　　　　　　　　　　　　　蒲　池　幹　治
　　　　　　　　　　　　　　　　　　発行者　秀　島　　　功
　　　　　　　　　　　　　　　　　　印刷者　横　山　明　弘

発行所　三共出版株式会社　　東京都千代田区神田神保町3の2
　　　　　　　　　　　　　　　　　　振替　00110-9-1065
郵便番号　101-0051　電話　03-3264-5711(代) FAX 03-3265-5149

　　一般社団法人日本書籍出版協会・一般社団法人自然科学書協会・工学書協会　会員

　　　　　　　　　　　　　　　　　　　　　印刷・製本　横山印刷

JCOPY〈(一社)出版者著作権管理機構　委託出版物〉
本書の無断複写は著作権法上での例外を除き禁じられています．複写される場合は，そのつど事前
に，(一社)出版者著作権管理機構（電話 03-5244-5088, FAX 03-5244-5089, e-mail:info@jcopy.or.jp）
の許諾を得てください．

ISBN 978-4-7827-0649-7